AF539701

TEXT BOOK
OF
ALGEBRA

DPH MATHEMATICS SERIES

TEXT BOOK OF ALGEBRA

(For B.A., B.Sc.)

By

A.K. Sharma

DISCOVERY PUBLISHING HOUSE
NEW DELHI-110002

First Published-2004
Reprinted-2010

ISBN 81-7141-896-1

Published by

DISCOVERY PUBLISHING HOUSE
4831/24, Ansari Road, Prahlad Street,
Darya Ganj, New Delhi-110002 (India)
Phone: 23279245 • Fax: 91-11-23253475
E-mail:dphtemp@indiatimes.com

Printed at :
Sachin Printers
Delhi

PREFACE

This book on Text Book of Algebra has been specially written to meet the requirements of the B.A. and B.Sc.

The subject matter has been discussed in such a simple way that the students will find no difficulty to understand it. The proofs of various theorems and examples have been given with minute details. Each chapter of this book contains complete theory and a fairly large number of solved examples. Sufficient problems have also been selected from various university examination papers. At the end of each chapter an exercise containing objective questions only has been given. The authors' shall be grateful to the readers who point out errors and omissions which, inspite of all care, might have been there.

The authors in general hope that the present book will be warmly received by the students and teachers. We shall indeed be very thankful to our colleagues for their recommending this book to their students.

The authors will feel amply rewarded if the book serves the purpose for which it is meant. Suggestions for the improvement of the book are always welcome.

—The Author

CONTENTS

1

RELATION BETWEEN THE ROOTS AND THE COEFFICIENTS OF AN EQUATION

1.1 INTRODUCTION

The general formed an algebraic equation is given as:

$$a_0x^4 + a_1x^{n-1} + \ldots, a_{n-1}x + a_n = 0$$

where $a_0 \neq 0$ and n is a positive integer. The term an which does not contain x is called *absolute term* or *constant term*.

An equation is said to be complex when all the power of variable x from x^n to x^0 are present in the equation. It any power of x is missing then it is said to incomplete equation. Any incomplete equation can be made complete by adding the missing term with zero coefficient.

An equation is said to be numerical if all its coefficients are numbers while it is said to be algebrical if its coefficients are algebriac symbol.

Degree of an Equation

By the degree of an equation is sent the highest power of the variable occuring in the equation.

Thus the equation

$$a_0x^n + a_2x^{n-2} + a_3x^{n-3} \ldots, + a_n = 0$$

is said to be an equation of degree n where $a_0 \neq 0$.

An equation of second degree is called *quadratic* equation.

An equation of third degree is called a *cubic* equation.

An equation of fourth degree is called a *biqubic* equation.

Dynthetic Division

To find the quotient and reminder when a polynomial is divided by a binomial.

Let $f(x) \equiv a_0x^n + a_1x^{n+1} + \ldots, a_n$ be a polynomial of degree n and let it be divided by the binomial $x - h$. It $Q = b_0x^{n-1} + b_1x^{n-1} + \ldots, + b_{n-1}$ be the quotient and R is remainder, then the coefficient of Q and R can be exhibited in the following manner.

h	a_0	a_1	a_2	a_3	$\ldots a_{n-1}$	a_n
		hb_0	hb_2	a_3	hb_{n-2}	hb_{n-1}
	b_0	b_1	b_2	b_3	b_{n-1}	R

1.2 RELATION BETWEEN THE ROOTS AND COEFFICIENTS OF AN EQUATION

Let $\alpha_1, \alpha_2, \ldots, \alpha_n$ be the roots of the algebric equation of degree n

$$a_0x^n + a_1x^{n-1} + a_2x^{n-2} + \ldots + a_{n-1}\, x + a_n = 0, \qquad (a_0 \neq 0).$$

$$\therefore a_0x^n + a_1x^{n-1} + a_2x^{n-2} + \ldots + a_{n-1}\, x + a_n$$

$$\equiv a_0 (x - \alpha_1) (x - \alpha_2) \ldots (x - \alpha_n).$$

$$\Rightarrow \quad x^2 + \frac{a_1}{a_0} x^{n-1} + \frac{a_2}{a_0} x^{n-2} + \ldots + \frac{a_{n-1}}{a_0} x + \frac{a_n}{a_0}$$

$$= x^n - (\alpha_1 + \alpha_2 + \ldots + \alpha_n) x^{n-1} + (\alpha_1\alpha_2 + \alpha_1\alpha_3 + \ldots) x^{n-2} + \ldots$$

$$+ (-1)^n \alpha_1 \alpha_2 \ldots \alpha_n$$

$$= x^n - \Sigma\alpha_1 x^{n-1} + \Sigma\alpha_1\alpha_2\, x^{n-2} - \Sigma\alpha_1\alpha_2\alpha_3 x^{n-3} + \ldots + (-1)^n \alpha_1 \alpha_2 \ldots \alpha_n.$$

Comparing the coefficients of similar powers of x, we get

sum of the all the roots

$$\sigma_1 \equiv \Sigma\alpha_1 = -(a_1/a_0),$$

sum of the product of the roots taken two at a time

$$\sigma_2 \equiv \Sigma\alpha_1\alpha_2 = -a_2/a_0,$$

sum of the product of the roots taken three at a time

$$\sigma_3 \equiv \Sigma\alpha_1\alpha_2\,\alpha_3 = -(a_3/a_0).$$

$$\vdots$$

product of all roots

$$\sigma_n \equiv \alpha_1\alpha_2 \ldots \alpha_n = (-1)^n \ (a_n/a_0).$$

These are the required relations between the roots and coefficients of the given equation.

Particular Case

The relations between the roots and coefficients of quadratic, cubic and biquadratic equations are given below:

(a) If α and β be the roots of $a_0x^3 + a_1x + a_2 = 0$, then we have

$$\alpha + \beta = \frac{-a_1}{a_0} \text{ and } \alpha\beta = \frac{a_2}{a_0}. \ (a_0 \neq 0).$$

(b) Similarly if α, β, γ be the roots of $a_0x^3 + a_1x^2 + a_2x + a_3 = 0$, then we have

$$\sigma_1 \equiv \alpha + \beta + \gamma = \frac{-a_1}{a_0},$$

$$\sigma_2 \equiv \alpha\beta + \beta\gamma + \gamma\alpha = \frac{a_2}{a_0},$$

$$\text{and } \sigma_3 \equiv \alpha\beta\gamma = -\frac{a_3}{a_0}, \ (a_0 \neq 0).$$

(c) If α, β, γ, δ be the roots of $a_0x^4 + a_1x^3 + a_2x^2 + a_3x + a_4 = 0$ then we have

$$\sigma_1 \equiv \alpha + \beta + \gamma + \delta = \frac{-a_1}{a_0},$$

$$\sigma_2 \equiv \alpha\beta + \alpha\gamma + \alpha\delta + \beta\gamma + \beta\delta + \gamma\delta = \frac{a_2}{a_0},$$

$$\Rightarrow \quad \sigma_2 \equiv (\alpha + \beta)(\gamma + \delta) + \alpha\beta + \gamma\delta = \frac{a_2}{a_0},$$

$$\sigma_2 \equiv \alpha\beta\gamma + \alpha\beta\delta + \alpha\gamma\delta + \beta\gamma\delta = -\frac{a_3}{a_0},$$

$$\Rightarrow \quad \sigma_3 \equiv \alpha\beta(\gamma + \delta) + \gamma\delta(\alpha + \beta) = -\frac{a_3}{a_0},$$

$$\text{and } \quad \sigma_4 \equiv \alpha\beta\gamma\delta = \frac{a_4}{a_0}. \ (a_0 \neq 0)$$

1.3 SPECIAL ROOTS OF A CUBIC EQUATION

1. *Three root of a cubic equation in A. P, may be taken as*

$$a - d, \ a, \ a + d.$$

We have $\sigma_1 \equiv a - d + a + a + d = 3a$,

$$\sigma_2 \equiv a(a - d) + a(a + d) + (a - d)(a + d) = 3a^2 - a^2,$$

$$\sigma_3 \equiv a(a - d)(a + d) = a(a^2 - a^2).$$

2. *Three roots of a cubic equation in G. P, may be taken as*

$a/r, a, ar.$

We have $\sigma_1 \equiv a/r + a + ar$, $\sigma_2 \equiv a\,(a/r + a + ar)$, $\sigma_3 \equiv a^3$.

3. *Three roots of a cubic equation in H.P. may be taken as*

$$\alpha, \beta, \gamma \text{ where } \beta = \frac{2\alpha\gamma}{\alpha + \gamma}.$$

Since α, β, γ are in H. P., so $1/\alpha, 1/\beta, 1/\gamma$ are in A. P.

$$\text{Thus } \frac{1}{\beta} = \frac{1/\alpha + 1/\gamma}{2} = \frac{\gamma + \alpha}{2\alpha\gamma} \text{ and so } \beta = \frac{2\alpha\gamma}{\alpha + \gamma}$$

Theorem:

(a) if the equation

$$a_0x^n + a_1x^{n-1} + ... + a_{n-1}x + a_n = 0,\ a_0 \neq 0$$

where $a_0, a_1, ..., a_n$ are real numbers, has a complex root $p + iq$ (p, q are real numbers, $q \neq 0$), then it also has a complex root $p - iq$.

Or

The Complex Roots are Always Occur in Conjugate Pairs

(b) *If $a_0x^n + a_1x^{n-1} + ... + a_{n-1}x + a_n = 0$, $a_0 \neq 0$ where $a_0, a_1, ..., a_n$ are rational numbers, has an irrational root $p + \sqrt{q}$ (p, q are rational, $q > 0$, q is not the square of a rational number), then it also has an irrational root $p\ \sqrt{q}$.*

Or

The Irrational Roots Always Occur in Pairs

Remarks:

1. *The complex roots, of an equation with complex number efficient, may not occur in conjugate pairs.*

For example, the roots of $x^2 - 7ix - 12 = 0$ are $3i$ and $4i$, which are not conjugate to each other.

2. *The irrational roots, of an equation with irrational coefficients, my not occur in pairs.*

For example, the roots of $x^2 - 3\sqrt{3}x + 6 = 0$ are $\sqrt{3}$ and $2\sqrt{3}$.

However $-\sqrt{3}$ and $-2\sqrt{3}$ are not the roots of the equation.

Example 1:

Solve the equation $x^3 - 13x^2 + 15x + 189 = 0$, given that one of the roots recedes another by 2 (or given that the difference between two of its roots is 2.

Solution:

We have $x^3 - 13x^2 + 15x + 189 = 0$

Let α, β, γ be the roots of the given equation.

Let $\alpha - \beta = 2$ so that $\alpha = \beta + 2$. ...(1)

We have (2) $\alpha + \beta + \gamma = 13$, (3) $\alpha\beta\gamma = -189$.

From (1) and (2), $2\beta + \gamma = 11$.

Putting $\alpha = \beta + 2$ and $\gamma = 11 - 2\beta$ in (3), we get

$$\beta(\beta + 2)(11 - 2\beta) = -189.$$

i.e. $2\beta^3 - 7\beta^2 - 22\beta - 189 = 0$. ...(4)

Since β is a root of the given equation,

$\therefore$ $\beta^3 - 13\beta^2 + 15\beta + 189 = 0$. ...(5)

From (4) and (5), we get $19\beta^2 - 52\beta - 567 = 0$.

The roots of this quadratic equation are 7, $-\dfrac{81}{19}$

We take $\beta = 7$, since $-81/19$ does not satisfy the given equation. From (1), $\alpha = 9$. From (2), $\gamma = -3$.

Hence 9, 7, –3 are the roots of the given equation.

Example 2:

Solve the equation

$$2x^4 - 15x^3 + 35x^2 - 30x + 8 = 0,$$

given that the product to two roots is equal to the product of the other two.

Solution:

We have $2x^4 - 15x^3 + 35x^2 - 30x + 8 = 0$

Let the roots of the given equation be α, β, γ, δ, where

$\alpha\beta + \gamma\delta$. ...(1)

$\therefore \sigma_1 \equiv \alpha + \beta + \gamma + \delta = 15/2$, ...(2)

$\sigma_2 \equiv (\alpha + \beta)(\gamma + \delta) + \alpha\beta + \gamma\delta = 35/2$, ...(3)

$\sigma_3 \equiv (\alpha + \beta)\gamma\delta + (\gamma + \delta)\alpha\beta = 15$, ...(4)

and $\sigma_3 \equiv \alpha\beta\gamma\delta + 4$. ...(5)

From (1) and (5), we get

$\alpha\beta + \gamma\delta = 2$. ...(6)

From (3) and (6), we get

$$(\alpha + \beta)(\gamma + \delta) = \frac{35}{2} - 2 - 2 = \frac{27}{2}. \qquad ...(7)$$

From (2) and (7), it follows that $\alpha + \beta$ and $\gamma + \delta$ are the roots of $2y^2 - 15y + 27 = 0 \Rightarrow (2y - 9)(-3) = 0 \Rightarrow y = 9/2$ or 3.

Then $\alpha + \beta = 9/2$ and $\gamma + \delta = 3$.

We have $(\alpha - \beta)^2 = (\alpha + \beta)^2 - 4\alpha\beta + \frac{81}{4} - 8 = \frac{49}{4}$.

Thus $\alpha - \beta = \frac{7}{2}$. Also $\alpha + \beta = \frac{9}{2} \Rightarrow \alpha = 4$ and $\alpha = \frac{1}{2}$.

Again $(\gamma - \delta)^2 = (\gamma + \delta)^2 - 4\gamma\delta = 9 - 8 = 1$.

Thus $\gamma - \delta = 1$. Also $\gamma + \delta = 3 \Rightarrow \gamma = 2$ and $\delta = 1$.

Hence the roots are 4, $\frac{1}{2}$, 2 and 1.

Example 3:

The product of two the roots of the equation

$$x^4 + x^3 - 16x^2 - 4x + 48 = 0$$

is 6. Solve the equation.

Solution:

We have $x^4 + x^3 - 16x^2 - 4x + 48 = 0$

Let α, β, γ and δ be the roots of the given equation.

$$\therefore \quad \alpha + \beta + \gamma + \delta = -1, \qquad ...(1)$$

$$(\alpha + \beta)(\gamma + \delta) + \alpha\beta + \gamma\delta = -16, \qquad ...(2)$$

$$\alpha\beta(\gamma + \delta) + \gamma\delta(\alpha + \beta) = 4, \qquad ...(3)$$

$$\alpha\beta\gamma\delta = 48. \qquad ...(4)$$

It is given that $\alpha\beta = 6$ and so $\gamma\delta = 8$, by (4)

From (2), $(\alpha + \beta)(\gamma + \delta) + 6 + 8 = -16$

$$\Rightarrow \quad (\alpha + \beta)(\gamma + \delta) = -30. \qquad ...(5)$$

From (1) and (5), it follows that $\alpha + \beta$ and $\gamma + \delta$ are the roots of

$$y^2 + y - 30 = 0 \Rightarrow y = -6 \Rightarrow 5.$$

We have $\alpha + \beta = 5$ and $\gamma + \delta = -6$. ...(6)

[Notice that $\alpha + \beta = -6$ and $\gamma + \delta = 5$ do not satisfy (3)].

Now $\quad (\alpha - \beta)^2 = (\alpha + \beta)^2 - 4\alpha\beta = 25 - 24 = 1 \Rightarrow \alpha - \beta = \pm 1$, and $(\gamma - \delta)^2 = (\gamma + \delta)^2 - 4\gamma\delta = 36 - 32 = 4 \Rightarrow \gamma - \delta = \pm 2.$

Let us take $\alpha - \beta = 1$ and $\gamma - \delta = 2.$ ...(7)

From (6) and (7), $\alpha = 3, \beta = 2, \gamma = -2, \gamma = -4.$

Hence the roots are 3, 2, – 2. and –4.

Example 4:

Solve the equation $3x^4 - 25x^3 + 50x^2 - 50x + 12 = 0$, given that the product of two of its roots is 2.

Solution:

We have $3x^4 - 25x^3 + 50x^2 - 50x + 12 = 0$

Let α, β, γ and δ be the roots of the given equation, where $\alpha\beta = 2.$

Now
$$\alpha + \beta + \gamma + \delta = 25/3, \qquad ...(1)$$
$$(\alpha + \beta)(\gamma + \delta) + \alpha\beta + \gamma\delta = 50/3, \qquad ...(2)$$
$$\alpha\beta(\gamma + \delta) = \gamma\delta(\alpha + \beta) = 50/3, \qquad ...(3)$$
$$\sigma_1 \equiv \alpha\beta\gamma\delta = 4. \qquad ...(4)$$

Putting $\alpha\beta = 2$ in (3), we get $\gamma\delta = 2$ and so by (2),

$$(\alpha + \beta)(\gamma + \delta) = \frac{50}{3} - 4 = \frac{38}{3}. \qquad ...(5)$$

From (1) and (5), it follows that $\alpha + \beta, \gamma + \delta$ are the roots of

$$x^2 - \frac{25}{3}x + \frac{38}{3} = 0$$

$$\Rightarrow \quad 3x^2 - 25x + 38 = 0$$

$$\Rightarrow \quad (x - 2)(3x - 19) = 0$$

Thus $x = 2, 19/3 \Rightarrow \alpha + \beta = 2$ and $\gamma + \delta = 19/3.$

[Notice that (3) is also satisfied if $\alpha + \beta = 19/3$ and $\gamma + \delta = 2$].

Now $\alpha\beta = 2$ and $\alpha + \beta = 2 \Rightarrow \alpha + \dfrac{2}{\alpha} = 2 \Rightarrow \alpha^2 - 2\alpha + 2 = 0.$

$$\therefore \quad \alpha = \frac{2 \pm \sqrt{4 - 8}}{2} = \frac{2 \pm 2i}{2} = 1 \pm i.$$

Again gd = 2 and $\gamma + \delta = \dfrac{19}{3} \Rightarrow 3\gamma^2 - 19\gamma + 6 = 0$

$$\Rightarrow \quad (3\gamma - 1)(\gamma - 6) = 0 \Rightarrow \gamma = 1/3, 6.$$

Hence $1 \pm i$, 1/3 and 6 and 6 are the required roots.

Example 5:

Solve the equation $x^4 - 8x^3 + 14x^2 + 8x - 15 = 0$, the roots being in A. P.

Solution:

We have $x^4 - 8x^3 + 14x^2 + 8x - 15 = 0$

Let the required roots be a – d, a, a + d, a + 2d. Then

$\sigma_1 \equiv a - d + a + a + d + a + 2d = 8 Þ d = 4 - 2a.$ ...(1)

$\sigma_1 \equiv (a + a - d)(a + d + a + 2d) + a(a - d) + (a + d)(a + 2d) = 14$...(2)

From (1) and (2), we obtain

$(2a - 4 + 2a)(2a + 12 - 6a) + a(a - 4 + 2a) + (a + 4 - 2a)$
$(a + 8 - 4a) = 14$

Simplifying, $a^2 - 4a + 3 = 0 \Rightarrow a = 1, 3.$

Substituting a = 1 in (1), we get d = 2.

Hence the required roots are 1 – 2, 1, 1 + 2, 1 + 4

$\Rightarrow$ –1, 1, 3, 5.

Note: We get the same roots on taking a = 3 and d = – 2.

Example 6:

Solve the equation

$x^4 - 8x^3 + 21x^2 - 20x + 5 = 0;$

given that the sum of two roots is equal to the sum of other two.

Solution:

We have $x^4 - 8x^3 + 21x^2 - 20x + 5 = 0$

Let α, β, γ, δ, be the roots of the given equation.

Let $\alpha + \beta = \gamma + \delta.$ ...(1)

We have $\sigma_1 \equiv \alpha + \beta + \gamma + \delta = 8,$ ...(2)

$\sigma_2 \equiv (\alpha + \beta)(\gamma + \delta) + \alpha\beta + \gamma\delta = 21,$ ...(3)

$(\alpha + \beta)\gamma\delta + (\gamma + \delta)\alpha\beta = 20$ and $\alpha\beta\gamma\delta = 5.$ ...(4)

(1) and (2) $\Rightarrow \alpha = \beta = \gamma + \delta = 4,$ ...(5)

(3) and (5) $\Rightarrow \alpha\beta + \gamma\delta = 5.$ ...(6)

Let $\alpha\beta = x$ and $\gamma\delta = y$ so that $x = y = 5$ and $xy = 5$ and $xy = 5$, (4)

Now $(x - y)^2 = (x + y)^2 - 4xy = 25 - 20 = 5$

$\Rightarrow \quad x - y = \pm \sqrt{5}$. Let $x - y = \sqrt{5}$.

Now $x + y = 5$ and $x - y = \sqrt{5}$

$$\Rightarrow \quad x = \frac{5+\sqrt{5}}{2},\ y = \frac{5-\sqrt{5}}{2},$$

$$\Rightarrow \quad \alpha\beta = \frac{1}{2}(5+\sqrt{5}) \text{ and } \gamma\delta = \frac{1}{2}(5-\sqrt{5}).$$

Now $(\alpha - \beta)^2 = (\alpha + \beta)^2 - 4\alpha\beta = 16 - 2(5 - \sqrt{5}) = 6 - 2\sqrt{5}$

$\Rightarrow \quad (\alpha - \beta)^2 = (\sqrt{5} - 1)^2$. We take $\alpha - \beta = \sqrt{5} - 1$.

Now $\alpha + \beta = 4$ and $\alpha - \beta = \sqrt{5} - 1$

$$\Rightarrow \quad \alpha = \frac{1}{2}(3+\sqrt{5}),$$

and $$\beta = \frac{1}{2}(5+\sqrt{5}).$$

Now $(\gamma - \delta)^2 = (\gamma + \delta)^2 - 4\gamma\delta = 16 - 2(5 - \sqrt{5}) = 6 + 2\sqrt{5}$

$\Rightarrow \quad (\gamma + \delta)^2 = (\sqrt{5} + 1)^2$. We take $\gamma - \delta\ \sqrt{5} + 1$.

Since $\gamma + \delta = 4$ and $\gamma - \delta = \sqrt{5} + 1$, $\therefore = \frac{1}{2}(5+\sqrt{5})$,

$$\delta = \frac{1}{2}(3+\sqrt{5}).$$

Hence the roots are $\frac{1}{2}(3 \pm \sqrt{5}),\ \frac{1}{2}(5 \pm \sqrt{5})$.

Example 7:

Solve the equation $x^4 - 2x^3 + 4x^2 + 6x - 21 = 0$, being given that it has two roots equal in magnitude but opposite in sign.

Solution:

We have $x^4 - 2x^3 + 4x^2 + 6x - 21 = 0$

Let α, β, γ and δ be the roots of the given equation where

$$\alpha = -\beta \Rightarrow \alpha + \beta\ 0. \quad ...(1)$$

We have $\alpha + \beta + \gamma + \delta = 2, \quad ...(2)$

$$(\alpha + \beta)(\gamma + \delta) + \alpha\beta + \gamma\delta = 4, \quad ...(3)$$

$$\alpha\beta(\gamma + \delta) + \gamma\delta(\alpha + \beta) = -6, \quad ...(4)$$

and $\alpha\beta\gamma\delta = -21. \quad ...(5)$

Using (1) in (2), (4); we get respectively

$$\gamma + \delta = 2,\ \alpha\beta(\gamma + \delta) = -6,$$

i.e., $\alpha\beta = -3, \gamma + \delta = 2$ and $\gamma\delta = 7$, by (5).

Now $(\alpha - \beta)^2 = (\alpha + \beta)^2 - 4\alpha\beta = 12 \Rightarrow \alpha - \beta = \pm 2\sqrt{3}$. ...(6)

$(\gamma - \delta)^2 = (\gamma + \delta)^2 - 4\gamma\delta = 4 - 28 = -24 \Rightarrow \gamma = \pm i\, 2\sqrt{6}$.

Adding (1) and (6), we get $a = \pm \sqrt{3}$.

Let $\gamma - \delta = i\, 2\sqrt{6}$.

$\therefore$ $(\gamma + \delta) + (\gamma - \delta) = 2 + i\, 2\sqrt{6} \Rightarrow \gamma = 1\ i\sqrt{6}$.

$(\gamma + \delta) - (\gamma - \delta) = 2 - i\, 2\sqrt{6} \Rightarrow \delta = 1\ i\sqrt{6}$.

Hence the roots are $\pm\sqrt{3}$, $1 \pm i\sqrt{6}$.

Example 8:

The product of two of the roots of the equation $x^4 - 5x^3 + 10x^2 - 10x + 4 = 0$ is equal to the product of the other two. Solve the equation.

Solution:

We have $x^4 - 5x^3 + 10x^2 - 10x + 4 = 0$

Let α, β, γ and δ be the roots of the given equation where

$\alpha\beta = \gamma\delta$. ...(1)

We have $\sigma_1 \equiv (\alpha + \beta) + (\gamma + \delta) = 5$, ...(2)

$\sigma_3 \equiv (\alpha + \beta)\gamma\delta + \alpha\beta(\gamma + \delta) = 10$, ...(3)

$\sigma_4 \equiv \alpha\beta\gamma\delta = (\alpha\beta)(\gamma\delta) = 4 \Rightarrow (\alpha\beta)^2 = 4$, by (1)

We take $\alpha\beta = \gamma\delta = 2$. ...(4)

Now $\sigma_2 \equiv (\alpha + \beta) + (\gamma + \delta) + \alpha\beta + \gamma\delta = 10$.

Using (4), $(\alpha + \beta)(\gamma + \delta) = 6$. ...(5)

From (2) and (5), $(\alpha + \beta)$ and $(\gamma + \delta)$ are the roots of the equation

$y^2 - 5y + 6 = 0 \Rightarrow y = 3, 2$.

Then $\alpha = \beta = 3$ and $\gamma + \delta = 2$. ...(6)

Now $(\alpha - \beta)^2 = (\alpha + \beta)^2 - 4\alpha\beta - 9 - 4(2) = 1 \Rightarrow -\beta = \pm 1$.

$(\gamma - \gamma)^2 = (\gamma + \delta)^2 - 4\gamma\delta = 4 - 4(2) = -4 \Rightarrow \gamma - \delta = \pm 2i$.

Let $\alpha - \beta = 1$ and $y - \delta = 2i$. ...(7)

From (6) and (7), $\alpha = 2$, $\beta = 1$; $\gamma = 1 + i$, $\delta = 1 - i$.

Hence the roots are 2, 1, $1 \pm i$.

Example 9:

If 1, α_1, α_2...α_{n-1} be the roots of the equation $x^n - 1 = 0$, show that $n = (1 - \alpha_1)(x - \alpha_2)...(1 - \alpha_{n-1})$.

Solution:

Since 1, $\alpha_1, \alpha_2,...\alpha_{n-1}$ be the roots of $x^n - 1 = 0$,

$\therefore x^n - 1 = (x - 1)(x - \alpha_1)(x - \alpha_2)...(x - \alpha_{n-1})$.

Dividing both sides by $(x - 1)$, we get

$$\frac{x^n - 1}{x - 1} = (x - \alpha_1)(x - \alpha_2)...(x - \alpha_{n-1})$$

$\Rightarrow \quad x^{n-1} + x^{n-2} + ... + x + 1 = (x - \alpha_1)(x - \alpha_2)...(x - \alpha_{n-1})$.

Putting $x = 1$, $n = (1 - \alpha_1)(1 - \alpha_2)...(1 - \alpha_{n-1})$.

Example 10:

The product of two roots of the equation

$$x^4 - 10x^3 + 42x^2 - 82x + 65 = 0$$

Solve the equation.

Solution:

We have $x^4 - 10x^3 + 42x^2 - 82x + 65 = 0$

Let $\alpha, \beta, \gamma, \delta$ be the roots of the given equation.

$\therefore \quad \alpha + \beta + \gamma + \delta = 10,$ ...(1)

$(\alpha + \beta)(\gamma + \delta) + \alpha\beta + \gamma\delta = 42,$ ...(2)

$\alpha\beta(\gamma + \delta) + \gamma\delta(\alpha + \beta) = 82,$ ...(3)

$\alpha\beta\gamma\delta = 65.$ ...(4)

Let $\alpha\beta = 13$. then $\gamma\delta = 5$, using (4).

From (2), $(\alpha + \beta)(\gamma + \delta) + 13 + 5 = 42 \Rightarrow (\alpha + \beta)(\gamma + \delta) = 24$ ...(5)

From (1) and (5), if follows that $\alpha + \beta$, $\gamma + \delta$ are the roots of

$$y^2 - 10y + 24 = 0 \Rightarrow (y - 6)(y - 4) = 0 \Rightarrow y = 4, 6.$$

$\Rightarrow \quad \alpha + \beta = 6$ and $\gamma + \delta = 4.$ ...(6)

[Notice that (3) is not satisfied on taking $\alpha + \beta = 4$ and $\gamma + \delta = 6$]

Now $(\alpha + \beta)^2 = (\alpha + \beta)^2 - 4\alpha\beta = 36 - 52 = -16 \Rightarrow \alpha - \beta = \pm 4i,$

$(\gamma - \delta)^2 = (\gamma + \delta)^2 - 4\gamma\delta = 16 - 20 = -4 \Rightarrow \gamma - \delta = \pm 2i.$

$\therefore \quad \alpha + \delta = 6, \alpha - \beta = \pm 4i \Rightarrow 3 \pm 2i$ are two roots,

$\gamma + \delta = 4, \gamma - \delta = \pm 2i \Rightarrow 2 \pm i$ are the other two roots.

Hence $3 \pm 2i, 2 \pm i$ are the required roots.

Example 11:

If $\alpha, \beta, \gamma, \delta$ be the roots of the equation

$$x^4 + px^3 + qx^2 + rx + s = 0, \text{ show that}$$

$$(1 + \alpha^2)(1 + \beta^2)(1 + \gamma^2)(1 + \delta^2) = (1 - q + a)^2 + (p - r)^2.$$

Solution:

Since $\alpha, \beta, \gamma, \delta$ are the roots of the given equation,

$\therefore x^4 + px^3 + qx^2 + rx + s = (x - \alpha)(x - \beta)(x - \gamma)(x - \delta)$. ...(1)

Putting $x = i$ and $-i$ in (1), we respectively obtain

$(1 - q + s) - i(p - r) = (i - \alpha)(i - \beta)(i - \gamma)(i - \delta)$, ...(2)

and $(1 - q + s) + i(p - r) = (-i - \alpha)(-i - \beta)(-i - \gamma)(-i - \delta)$. ...(3)

Multiplying both sides of (2) and (3), we get

$$(1 - q + s)^2 + (p - r)^2 = (1 + \alpha^2)(1 + \beta^2)(1 + \gamma^2)(1 + \delta^2).$$

Example 12:

If α, β, γ are the roots of $x^3 - ax^2 + bx - c = 0$ evaluate

(i) $\Sigma\left(\frac{\alpha}{\beta} + \frac{\beta}{\alpha}\right)$, (ii) $\Sigma \alpha^3$.

Solution:

We have

$\alpha + \beta + \gamma = a, \alpha\beta + \beta\gamma + \gamma\alpha = \beta, \alpha\beta\gamma = c.$ ...(1)

(i) $\Sigma\left(\frac{\alpha}{\beta} + \frac{\beta}{\alpha}\right) = \Sigma\frac{\alpha^2 + \beta^2}{\alpha\beta} = \Sigma\frac{(\alpha^2 + \beta^2)\gamma}{\alpha\beta\gamma}$

$= \frac{1}{c}[(\alpha^2\gamma + \beta^2\gamma) + (\beta^2\alpha + \gamma^2\alpha) + (\gamma^2\beta + \alpha^2\beta)$

$= \frac{1}{c}\Sigma\alpha^2\beta.$ ...(2)

Consider $\Sigma\alpha\,\Sigma\alpha\beta = \Sigma\alpha^2\beta + 3\alpha\beta\gamma$.

Using (1), $\Sigma\alpha^2\beta = ab - 3c$, ...(3)

Putting (3) in (2), $\Sigma\left(\frac{\alpha}{\beta} + \frac{\beta}{\alpha}\right) = \frac{ab - 3c}{c}$.

(ii) $(\alpha + \beta + \gamma)^2 = \Sigma\alpha^2 + 2\Sigma\alpha\beta$

$\Rightarrow \quad \Sigma\alpha^2 = a^2 - 2b.$ ...(4)

We have $\Sigma\alpha\Sigma x^2 = (\alpha + \beta + \gamma)(\alpha^2 + \beta^2 + \gamma^2)$

$\Rightarrow \quad \Sigma x\Sigma\alpha^2 = \Sigma\alpha^3 + \Sigma\alpha^2\beta.$

$\therefore \Sigma\alpha^3 = a(a^2 - 2b) - (ab - 3c)$, using (3) and (4).

Hence $\Sigma\alpha^3 = a^3 - 3ab + 3c.$

Example 13:

Solve the equation $x^3 - 9x^2 + 23x - 15 = 0$, the roots being in A.P.

Solution:

We have $x^3 - 9x^2 + 23x - 15 = 0$,

Let $a - d$, a $a + d$ be the roots of the given equation. Then $a - d$, $a + a + d = 9 \Rightarrow a = 3.$

$$a(a - d) + a(a + d) + (a - d)(a + d) = 23$$

$\Rightarrow \quad 2a^2 + a^2 - d^2 = 23 \Rightarrow 3a^2 - d^2 = 23$

$\Rightarrow \quad d^2 = 4 \ (\because a = 3) \Rightarrow d = \pm 2.$

Hence the required roots are $3 - 2, 3, 3 + 2$ (or $3 + 2, 3, 3 - 2$) *i.e.*, 1, 3, 5.

Example 14:

Solve the equation $x^3 = -9x^2 + 23x - 15 = 0$, two of the roots being in the ration 3 : 5.

Solution:

We have $x^3 = -9x^2 + 23x - 15 = 0$

Let 3α, 5α and β be the roots of the given equation.

$\therefore 3\alpha + 5\alpha + \beta = 9$ and $\sigma_2 \equiv 15a^2 + 8\alpha\beta = 23.$ ...(1)

From (1), $15\alpha^2 + 8\alpha(9 - 8\alpha) = 23$ or $49\alpha^2 - 27\alpha + 23 = 0.$

$$\Rightarrow \quad (49\alpha - 23)(\alpha - 1) = 0. \quad \therefore \quad \alpha = 1 \Rightarrow \frac{23}{49}.$$

We see that $\alpha = 23/49$ does not satisfy the given equation. Taking $\alpha = 1$ in (1), we get $\beta = 1.$

Hence the roots of the given equation are 3, 5 and 1.

Example 15:

Solve the equation $2x^3 - x^2 - 22x - 24 = 0$, two of the roots being in the ration 3 : 4.

Solution:

We have $2x^3 - x^2 - 22x - 24 = 0$

Let the roots be 3α, 4α, β. Then

$$3\alpha + 4\alpha + \beta = \frac{1}{2} \Rightarrow \beta = \frac{1}{2} - 7a, \qquad ...(1)$$

and $3\alpha, 4\alpha, + 3\alpha.\beta + 4\alpha.\beta = -11 \Rightarrow 12\alpha^2 + 7\alpha\beta = -11$ ...(2)

From (1) and (2),

$$22\alpha^2 + 7\alpha\left(\frac{1}{2} - 7\alpha\right) = -11 \Rightarrow 74\alpha^2 - 7\alpha - 22 = 0$$

$$\therefore \alpha = \frac{7 \pm \sqrt{49 + 6512}}{148} = \frac{7 \pm 81}{148} = -\frac{1}{2}, \frac{22}{37}.$$

We take $\alpha = -1/2$, since $\alpha = 22/37$ does not satisfy the given equation. From (1), $\beta = 4$.

Hence the required roots are $-3/2, -2, 4$.

Example 16:

Solve the equation $x^3 - 5x^2 - 16x + 80 = 0$, then sum of two of its roots being zero.

Solution:

Let the three root be α, β, γ where $\alpha + \beta = 0$.

We have $\alpha + \beta + \gamma = 5$ and so $\gamma = 5$.

Again $\alpha\beta\gamma = -80 \Rightarrow \alpha(-a)(5) = -80 \Rightarrow \alpha^2 = 16 \Rightarrow \alpha = \pm 4$.

Hence the roots are $4, -4, 5$.

Example 17:

Solve the cubic equation $2x^3 - 9x^2 + 12x - 4 = 0$, given that two of its roots are equal.

Solution:

We have $2x^3 - 9x^2 + 12x - 4 = 0$

Let the three roots be α, α, γ. Then

$$\sigma_1 \equiv \alpha + \alpha + \gamma = \frac{9}{2} \Rightarrow 2a + \gamma = \frac{9}{2} \Rightarrow \gamma = \frac{9}{2} - 2\alpha. \qquad ...(1)$$

$$\sigma_2 \equiv \alpha.\alpha + \alpha.\ \gamma + \alpha.\gamma = 12/2$$

$$\Rightarrow \alpha^2 + 2\alpha\gamma = 6. \qquad ...(2)$$

From (1) and (2), $\alpha^2 + 2\alpha\left(\frac{9}{2} - 2\alpha\right) = 6 \Rightarrow \alpha^2 - 3\alpha + 2 = 0$

or, $(\alpha - 2)(\alpha - 1) = 0 \Rightarrow \alpha = 1.2$ Since $\alpha = 1$ does not satisfy the given equation, so $\alpha = 2$. From (1), $\gamma = \frac{1}{2}$. Hence the required roots are 2, 2, $\frac{1}{2}$

Example 18:

Find the equation of the lowest degree with rational coefficients having $2 + \sqrt{3}$ *and* $\sqrt{5} - 2$ *as two of it roots.*

Solution:

Since irrational roots occur in pairs, the required equation must have at least four roots viz. $2 + \sqrt{3}, 2 - \sqrt{3}, -2 + \sqrt{5} -2 - \sqrt{5}$. The required equation is

$$\{x - (2 + \sqrt{3})\}\ (x - (2 - \sqrt{3})\}\{x - (2 + \sqrt{5})\}\ \{(-2 - \sqrt{5})\} = 0$$

$$\Rightarrow \quad \{(x - 2) - \sqrt{3}\}\ \{x - 2) + \sqrt{3}\}\ \{(x + 2 - \sqrt{5}\}\ \{(x + 2) + \sqrt{5}\} = 0$$

Hence $\{(x - 2)^2 - 3\}\{x + 2)^2 - 5\} = 0$ is the required equation.

Example 19:

Form the equation of the lowest degree with real coefficients with has $1 + 3i$ *and* $2 - i$ *as tow of its roots.*

Solution:

Since complex roots occur in conjugate paris, the required equation must have at least four roots *viz.*, $1 + 3i$, $1 + 3i$, $2 - i$ and $2 - i$. The required equation is

$$\{x - (1 + 3i)\}\ \{x - (1 - 3i)\}\ \{x - (2 - i)\}\ \{x - (2 + i)\} = 0$$

$$\Rightarrow \quad \{(x - 1) - 3i\}\ \{(x - 1) + 3i\}\ \{(x - 2) + i\} + i\ \{(x - 2) - i\} = 0$$

$$\Rightarrow \quad \{(x - 1)^2 - (3i)^2\}\ \{(x - 2)^2 - i^2\} = 0.$$

Hence $\{(x - 1)^2 + 9\}\ \{(x - 2)^2 + 1\} = 0$ is the required equation.

Example 20:

Find a necessary condition for the sum of two roots of $x^4 - px^3 + qx^3 + qx^2 - rx + s = 0$ *to be equal to the sum of the other two.*

Solution:

We have $x^4 - px^3 + qx^3 + qx^2 - rx + s = 0$

Let the roots of the given equation be $\alpha, \beta, \gamma, \delta$,

where $\alpha + \beta = \gamma + \delta.$ ...(1)

We have $\sigma_1 \equiv \alpha + \beta + \gamma + \delta = p$. So by (1), $\alpha + \beta = \gamma + \delta = p/2$. ...(2)

Now $\sigma_2 \equiv (\alpha + \beta)(\gamma + \delta) + \alpha\beta + \gamma\delta = q.$ ...(3)

$\sigma_3 \equiv \alpha\beta(\alpha + \beta)\, \gamma\delta(\alpha + \delta) = r.$ ...(4)

Putting (2) in (3) and (4), we respectively obtain

$\alpha\beta + \gamma\delta = q - (p^2/4)$ and $\alpha\beta + \gamma\delta = 2r/p.$

Hence $q - \dfrac{p^2}{4} = \dfrac{2r}{p} \Rightarrow p^3 - apq + 8r = 0$

is the required condition.

Example 21:

Find a necessary condition for the roots of the equation $a_0x^3 + a_1x^2 + a_2x + a_3 = 0$ to be in G.P.

Solution:

We have $a_0x^3 + a_1x^2 + a_2x + a_3 = 0$

Let roots be $\alpha/\beta, \alpha, \alpha\beta$. Then

$$\sigma_1 \equiv \frac{\alpha}{\beta} + \alpha + \alpha\beta = -\frac{a_1}{a_0}, \quad ...(1)$$

$$\sigma_1 \equiv \alpha\left(\frac{\alpha}{\beta} + \alpha + \alpha\beta\right) = \frac{a_1}{a_0}, \quad ...(2)$$

$$\sigma_1 \equiv \alpha^3 = -\frac{a_1}{a_0}. \quad ...(3)$$

Dividing (2) by (1), we get $\alpha = -\alpha_2/\alpha_1$.

Substituting in (3), $a_0a_2^3 = a_3a_1^3$ is the required condition.

Example 22:

Find a necessary condition for the roots of the equation $x^3 - px^2 + qx - r = 0$ to be in

(1) A.P. (2) G.P. (3) H.P.

Solution:

1. Let $\alpha - \beta$, α, $\alpha + \beta$ be the roots of given equation.

Then $(\alpha - \beta) + \alpha\ (\alpha + \beta) = p \Rightarrow \alpha = p/3$. ...(1)

Since α is a root of the given equation,

$\therefore \alpha^3 - p\alpha^2 + q\alpha - r = 0$. ...(2)

Substituting the value of a from (1) in (2), we see that

$2p^3 + 9pq + 27\ r = 0$ is the required condition.

2. Let the roots be α/β, α, $\alpha\beta$. Then

$$\sigma_1 \equiv \frac{\alpha}{\beta} + \alpha + \alpha\beta = p, \qquad ...(3)$$

$$\sigma_1 \equiv \alpha\left(\frac{\alpha}{\beta} + \alpha + \alpha\beta\right) = q, \qquad ...(4)$$

$$\sigma_1 \equiv \alpha^3 = r. \qquad ...(5)$$

Dividing (4) by (3), $\alpha = q/p$.

Substituting in (5), the required condition is $q^3 = p^3 r$.

3. Let the roots be α, β, γ. Then $\beta = \dfrac{2\alpha\gamma}{\alpha + \gamma}$.

$\Rightarrow \quad \alpha\beta + \beta\gamma + \gamma\alpha = 3\alpha\gamma \Rightarrow q = 3\ \alpha\gamma \Rightarrow \alpha\gamma = q/3$. ...(6)

Since $\alpha\beta\gamma = r$, so $\beta = 3r/q$, using (6)

Since β is a root of the given equation,

$\therefore \beta^3 - p\beta^2 + q\beta - r = 0$. Putting $\beta = 3r/q$, we see that

$27r^3 - 9pqr^2 + 2q^3r = 0$ is the required condition.

Example 23:

Solve the equation $28x^3 - 39x^2 + 12x - 1 = 0$, the roots being in H.P.

Solution:

We have $28x^3 - 39x^2 + 12x - 1 = 0$

Let α, β, γ the roots of the given equation so that

$$\beta = \frac{2\alpha\gamma}{\alpha + \gamma}.$$

$\Rightarrow \quad \alpha\beta + \beta\gamma + \gamma\alpha = 3\alpha\gamma$. ...(1)

Now $\alpha\beta + \beta\gamma + \gamma\alpha = 12/28 = 3/7$. ...(2)

From (1) and (2), $\alpha\gamma = 1/17$. also $\alpha\beta\gamma = 1/28$. Thus $\beta = 1/4$.

Thus 1/4 is a root of the given equation or $(x - 1/4)$ or $(4x - 1)$ is a factor of the given equation.

Dividing $28x^3 - 39x^2 + 12x - 1$ by $(4x - 1)$, we obtain

$$(4x - 1)(7x^2 - 8x + 1) = 0 \Rightarrow (4x - 1)(7x - 1)(x - 1) = 0.$$

Hence 1, 1/4, 1/7 are the required roots.

Example 24:

Solve the equation $8x^3 - 14x^2 + 7x - 1 = 0$, the roots being in G. P.

Solution:

We have $8x^3 - 14x^2 + 7x - 1 = 0$

Let α/β, α $\alpha\beta$ be the roots of the given equation,

$$\therefore \quad \frac{\alpha}{\beta} + \alpha + \alpha\beta = \frac{7}{4} \qquad ...(1)$$

$$\sigma_1 \equiv \left(\frac{\alpha}{\beta}\right).\alpha.(\alpha\beta) = \frac{1}{8} \Rightarrow \alpha = \frac{1}{8} \Rightarrow \alpha = \frac{1}{2} \qquad ...(2)$$

From (1) and (2), we have $\frac{1}{2}\left(\frac{1}{\beta} + 1 + \beta\right) = \frac{7}{4}$

$\Rightarrow \quad 2\beta^2 - 5\beta + 2 = 0$

$\Rightarrow \quad (2\beta - 1)(\beta - 2) = 0 \Rightarrow \beta = 1/2 \Rightarrow 2.$

Hence the roots are $1, \frac{1}{2}, \frac{1}{4}$.

Example 25:

Solve the equation $2x^3 - 7x^2 + 7x - 2 = 0$, the roots bing in G. P.

Solution:

We have $2x^3 - 7x^2 + 7x - 2 = 0$

Let the roots be α/β, α and $\alpha\beta$.

Then $\quad \sigma_3 \equiv \alpha^3 = 1 \Rightarrow \alpha = 1.$

Also $\alpha/\beta + \alpha + \alpha\beta = 7/2$

$\Rightarrow \quad 2\beta^2 - 5\beta + 2 = 0 \qquad (\because a = 1).$

Thus $(\beta - 2)(2\beta - 1) = 0$

$\Rightarrow \quad \beta = 2, 1/2.$

Example 26:

Solve the equation $4x^4 + 8x^3 + 13x^2 + 2x + 3 = 0$, given that sum of two of the two roots is zero.

Solution:

We have $4x^4 + 8x^3 + 13x^2 + 2x + 3 = 0$

Let the roots be $\alpha, \beta, \gamma, \delta$, where $\alpha + \beta = 0$. ...(1)

We have $\sigma_1 \equiv \alpha + \beta + \gamma + \delta = -2,$ ...(2)

$\sigma_2 \equiv (\alpha + \beta)(\gamma + \delta) + \alpha\beta + \gamma\delta = 13/4,$ ...(3)

$\sigma_3 \equiv (\alpha + \beta)\gamma\delta(\gamma + \delta)\alpha\beta = -1/2,$ ...(4)

$\sigma_4 \equiv \alpha\beta\gamma\delta = 3/4.$ ...(5)

From (1) and (2), $\gamma + \delta = -2.$ ...(6)

From (1), (4) and (6); $-2\alpha\beta = -1/2$ or $\alpha\beta = 1/4.$ ...(7)

From (1) and (7), $\alpha^2 = -1/4 \Rightarrow = \pm \frac{1}{2} i.$

Dividing (5) and by (7), $\gamma\delta = 3 \Rightarrow \gamma(-2 - \gamma) = 3$, by (6)

$$\Rightarrow \quad \gamma^2 + 2\gamma + 3 = 0 \Rightarrow \gamma = \frac{-2 \pm \sqrt{4 - 12}}{2} = -1 \pm \sqrt{2}i.$$

Hence the roots are $\pm i/2, -1 \pm \sqrt{2}i.$

Example 27:

Solve the equation $x^4 + 2x^3 - 21x^2 - 22x + 40 = 0$, the sum of two of the roots being equal to the sum of the other two.

Solution:

We have $x^4 + 2x^3 - 21x^2 - 22x + 40 = 0$

Let $\alpha, \beta, \gamma, \delta$ be the roots of the given equation.

It is given that $\alpha + \beta = \gamma + \delta.$ ...(1)

We have $\sigma_1 \equiv \alpha + \beta + \gamma + \delta = -2,$ ...(2)

$\sigma_2 \equiv (\alpha + \beta)(\gamma + \delta) + \alpha\beta + \gamma\delta = -21,$ ...(3)

$\sigma_3 \equiv (\alpha + \beta)\gamma\delta(\gamma + \delta)\alpha\beta = 22,$ ...(4)

$\sigma_4 \equiv \alpha\beta\gamma\delta = 40.$ ...(5)

From (1) and (2), we obtain

$\alpha + \beta = \gamma + \delta = 1,$ ...(6)

and so be (3) $\alpha\beta + \gamma\delta = -22$. ...(7)

Let $\alpha\beta = x$ and $\gamma\delta = y$. Then by (7) and (5), we get

$$x + y = -22 \text{ and } xy = 40.$$

Now $(x - y)^2 = (x + y)^2 - 4xy = (-22)^2 - 4(40) = 324 = (18)^2$.

Thus $x - y = 18$ and $x + y = -22$

$\Rightarrow \quad x = -2, y = -20$

$\Rightarrow \quad \alpha\beta = -2$ and $\gamma\delta = -20$.

Now $(\alpha - \beta)^2 = (\alpha + \beta)^2 - 4\alpha\beta = (-1)^2 - 4(-2) = 9$, by (6)

Thus $\alpha - \beta = 3$ and $\alpha + \beta = -1 \Rightarrow \alpha = 1$ and $\beta = -2$.

Again $(\gamma - \delta)^2 = (\gamma + \delta)^2 - 4\gamma\delta = (-1)^2 - 4(-20) = 81$, by (6)

Thus $\gamma - \delta = 9$ and $\gamma + \delta = -1 \Rightarrow \gamma = 4$ and $\gamma - 5$

Hence the roots are 1, –2, 4, 5.

Example 28:

Solve the equation $2x^3 + x^2 - 7x - 6 = 0$, given that the difference of two of its roots is 3.

Solution:

We have $2x^3 + x^2 - 7x - 6 = 0$

Let α, β, γ be the roots of the given equation.

Let $\alpha - \beta = 3$ so that $\alpha = \beta + 3$. ...(1)

We have (2) $\alpha + \beta + \gamma - \frac{1}{2}$, (3) $\alpha\beta\gamma = 3$.

From (1) and (2), $\gamma = -(7/2) - 2\beta$. (4)

Putting (1) and (4) is (3), we get

$$\beta(\beta + 3)\left(\frac{7}{2} + 2\beta\right) = -3$$

$\Rightarrow \quad 4\beta^3 + 19\beta^2 + 21\beta + 6 = 0.$ (5)

Since β is a root of the given equation,

$2\beta^3 + \beta^2 - 7\beta - 6 = 0.$ (6)

Adding (5) and (6), we get

$6\beta^3 + 20\beta^2 + 14\beta = 0$

$\Rightarrow \quad 2\beta(3\beta^2 + 10\beta + 7) = 0$

$\Rightarrow$ $2\beta(3\beta + 7)(\beta + 1) = 0$

$\Rightarrow$ $\beta = 0, -1, -7/3$

It may be verified that only $\beta = -1$ satisfies (6), so

$\beta = -1, \alpha = 2$ and $\gamma = -3/2$, by (1) and 3)

Hence 2, –1, –3/2 are the required roots.

1.4 SYMMETRIC FUNCTIONS OF THE ROOTS

Let α, β, γ be the roots of a cubic equation.

The expressions

$$\alpha^2 + \beta^2 + \gamma^2, \alpha^2\beta^2 + \beta^2\gamma^2 + \gamma^2\alpha^2, (\alpha + \beta)(\beta + \gamma)(\gamma + \alpha) \text{ etc.}$$

are such that if any two of α, β, γ be interchanged, the above expressions remain unchanged. Such functions are called symmetric functions. In the following examples, we shall express symmetric functions of the roots of an equation in terms of the coefficients of the equation.

Example 1

If $\alpha + \beta + \gamma = 1$, $\alpha^2 + \beta^2 + \gamma^2 = 2$. $\alpha^3 + \beta^3 + \gamma^3 = 3$, find the value of $\alpha^4 + \beta^4 + \gamma^4$.

Solution:

Let the cubic equation with roots α, β, γ be

$$x^3 + px^2 + qx + r = 0.$$

Then $\alpha^3 + p\alpha^2 + q\alpha + r = 0$, $\beta^2 + p\beta^2 + q\beta + r = 0$, $\gamma^3 + p\gamma^2 + q\gamma = r = 0$.

Adding these,

$$(\alpha^3 + \beta^3 + \gamma^3) + p(\alpha^2 + \beta^2 + \gamma^2) + q(\alpha + \beta + \gamma) + 3r = 0. \quad \text{(A)}$$

$\Rightarrow$ $3 + 2p + q + 3r = 0$, as given. ...(1)

We have $\alpha + \beta + \gamma = -p$, $\alpha\beta + \beta\gamma + \gamma\alpha = q$ and $\alpha\beta\gamma = -r$.

Since $\alpha + \beta + \gamma = 1$, so $p = -1$. Now

$\alpha^2 + \beta^2 + \gamma^2 = (\alpha + \beta + \gamma)^2 - 2(\alpha\beta + \beta\gamma + \gamma\alpha)$

$\Rightarrow 2 = 1 - 2q \Rightarrow = -1/2$.

Putting $p = -1$ and $q = -1/2$ in (1), we get $r = -1/6$.

Thus the required equation is $x^3 + x^2 - \frac{1}{2}x - \frac{1}{6} = 0$

Multiplying the above equation by x, we get

$$x^4 - x^3 - \frac{1}{2}x^2 - \frac{1}{6}\ x = 0. \qquad ...(2)$$

Since α, β, γ satisfy (2), substituting α, β, γ for x in (2) and adding, we get

$$(\alpha^4 + \beta^4 + \gamma^4) - (\alpha^3 + \beta^3 + \gamma^3) - \frac{1}{2}(\alpha^2 + \beta^2 + \gamma^2) - \frac{1}{6}(\alpha + \beta + \gamma) = 0$$

Hence, $\alpha^4 + \beta^4 + \gamma^4 = 3 + \frac{1}{2}(2) + \frac{1}{6}(1) = \frac{25}{6}$.

Example 2:

If α, β, γ are the roots of $x^3 + qx - r = 0$, find the values of (1) $\sum \frac{\beta + \gamma}{\alpha^2}$ (2) Σa^4.

We have (1) $\alpha + \beta + \gamma = 0$, (2) $\Sigma\alpha\beta = q$, (3) $\alpha\beta\gamma = r$.

1. $\alpha + \beta + \gamma = 0 \Rightarrow \beta + \gamma = -\alpha$ and so

$$\sum \frac{\beta + \gamma}{\alpha^2} = -\sum \frac{\alpha}{\alpha^2} = -\sum \frac{1}{\alpha} = -\left(\frac{1}{\alpha} + \frac{1}{\beta} + \frac{1}{g}\right)$$

$$= -\frac{\sum \alpha\beta}{\alpha\beta\gamma} = -\frac{q}{r}.$$

2. $(\alpha + \beta + \gamma)^2 = \Sigma\alpha^2\beta + \Sigma\alpha\beta \Rightarrow \Sigma\alpha^2\alpha = -2q$, by (1) and (2).

Consider $\Sigma\alpha^2\ \Sigma\alpha^2 = (\alpha^2 + \beta^2 + \gamma^2)^2$

$\Rightarrow \quad (-2q)(-2q) = \Sigma\alpha^4 + 2\ \Sigma\alpha^2\beta^2$

$\therefore \quad \Sigma\alpha^4 = 4q^2 - 2\Sigma\alpha^2\beta^2. \qquad ...(4)$

Now $(\alpha\beta + \beta\gamma + \gamma\alpha)^2 = \Sigma\alpha^2\beta^2 + 2\alpha\beta\gamma(\alpha + \beta + \beta)$

$\Rightarrow \quad q^2 = \Sigma\alpha^2\beta^2 + 0 = \Sigma\alpha^2\beta^2$, by (1) and (2).

Putting in (4), $\Sigma\alpha^4 = 2q^2$.

Example 3:

If $\alpha + \beta + \gamma = 0$, show that

$$3(\alpha^2 + \beta^2 + \gamma^2)(\alpha^5 + \beta^5 + \gamma^5) = 5(\alpha^3 + \beta^3 + \gamma^3)(\alpha^4 + \beta^4 + \gamma^4).$$

Solution:

Since $\alpha + \beta + \gamma = 0$, the equation with roots α, β, and γ may be taken as $x^3 + qx + r = 0$.

From equations (3), (4), (6) and **(8)**

$\alpha^2 + \beta^2 + \gamma^2 = -2q, \quad \alpha^3 + \beta^3 + \gamma^3 = -3r,$

$\alpha^4 + \beta^4 + \gamma^4 = 2q^2$, $\alpha^5 + \beta^5 + \gamma^5 = 5qr$.

L.H.S. $= 3(\alpha^2 + \beta^2 + \gamma^2)(\alpha^5 + \beta^5 + \gamma^5) = 3(-2q)(5qr) = -30q^2r$.

R.H.S. $= 5(\alpha^3 + \beta^3 + \gamma^3)(\alpha^4 + \beta^4 + \gamma^4) = 5(-3r)(2q^2) = -30q^2r$.

Hence the result.

Example 4:

If a, b, g are the roots of $x^3 + qx + r = 0$, then

$$\frac{\alpha^5+\beta^5+\gamma^5}{5} = \frac{\alpha^2+\beta^2+\gamma^2}{2}\cdot\frac{\alpha^3+\beta^3+\gamma^3}{3}$$

Solution:

We have $\dfrac{\alpha^5+\beta^5+\gamma^5}{5} = \dfrac{\alpha^2+\beta^2+\gamma^2}{2}\cdot\dfrac{\alpha^3+\beta^3+\gamma^3}{3}$

$$\frac{\alpha^5+\beta^5+\gamma^5}{5} = qr = \frac{\alpha^2+\beta^2+\gamma^2}{-2}\cdot\frac{\alpha^3+\beta^3+\gamma^3}{-3},$$

Hence $\dfrac{\alpha^5+\beta^5+\gamma^5}{5} = \dfrac{\alpha^2+\beta^2+\gamma^2}{2}\cdot\dfrac{\alpha^3+\beta^3+\gamma^3}{3}$.

Example 5:

Find the sum of the cubes of the roots of the equation $x^3 - 6x^2 + 11x - 6 = 0$.

Solution:

We have $x^3 - 6x^2 + 11x - 6 = 0$

Let $f(x) = x^3 - 6x^2 + 11x - 6$. $\therefore f'(x) = 3x^2 - 12x + 11$.

Let α, β, γ be the roots of the given equation.

Then $\alpha^3 + \beta^3 + \gamma^3$ is the coefficient of $1/x^4$ in the expansion of $\dfrac{f'(x)}{f(x)}$.

We now divide $f'(x)$ by $f(x)$ till $\dfrac{1}{x^4}$ appears in the quotient.

$$3/x + 6/x^2 + 14/x^2 + 36/x^4 \qquad ...(1)$$

$x^3 - 6x^2 + 11x + 6 \quad 3x^3 - 12x + 11$

$x^2 - 18x + 33 - 18/x$

$6x - 22 + 18/x$

$6x - 36 + 66/x - 36/x^2$

$14 - 48/x + 36/x^2$

$$14 - 84/x \ 154/x^2 - 84/x^3$$
$$36/x - 118/x^2 + 84/x^3$$
$$36/x - \ldots\ldots\ldots\ldots$$

The coefficient of $1/x^4$ is 36, from (1).

Hence $\alpha^3 + \beta^2 + \gamma^2 = 36$.

Example 6:

Find the sum of the reciprocals of the fifth powers of α, β, γ – the roots of the equation $x^3 + 2x^2 + 1 = 0$.

Solution:

Let $f(x) = x^3 + 2x^3 + 1$, then $f'(x) = 3x^2 + 4x$.

Arranging f' (x), f (x) in assecending powers of x, we get

$$\frac{f'(x)}{f(x)} = \frac{(-4x - 3x^2)}{1 + 2x^2 + x^3}$$
$$= (-4x - 3x^2)\,[1 + (2x^2 + x^3)]^{-1}$$
$$= (-4x - 3x^2)\,[1 + (2x^2 + x^3) + (2x^2 + x^3)^2 + \ldots]$$
$$= (-4x - 3x^2) - (-4x - 3x^2)\,(2x^2 + x^3) + \ldots \qquad \ldots(1)$$

$\therefore \alpha^{-5} + \beta^{-5} + \gamma^{-5}$ = coefficient of x^4 in (1)

$$= -(-4 - 6) = 10$$

Hence $\alpha^{-5} + \beta^{-5} + \gamma^{-5} = 10$.

Example 7:

Find the value of $\alpha^{4} + \beta^{4} + \gamma^{4}$, given that α, β, γ are the roots of $x^2 - 7x + 7 = 0$.

Solution:

Let $f(x) = x^3 - 7x + 7$. Then $f'(x) = 3x^2 - 7$.

Arranging f' (x), f (x) in ascending powers of x, we get

$$\frac{f'(x)}{f(x)} = \frac{7 - 3x^2}{7 - 7x + x}$$
$$= \frac{7}{7}\left(1 - \frac{3}{7}x^2\right)\left[1 - \left(\frac{7x - x^3}{7}\right)\right]^{-1}$$

$$= \left(1 - \frac{3}{7}x^2\right)\left[1 + \left(\frac{7x - x^3}{7}\right) + \left(\frac{7x - x^3}{7}\right)^2 + \left(\frac{7x - x^3}{7}\right)^2 + \ldots\right] \quad \ldots(1)$$

Example 8:

If α, β, γ are the roots of $x^3 - px^2 + qx - r = 0$, find the values of (1) $\Sigma\alpha^2$, (2) $\Sigma\alpha^2\beta^2$, (3) $(\alpha + \beta)(\beta + \gamma)(\gamma + \alpha)$.)

Solution:

We have $x^3 - px^2 + qx - r = 0$

$$\alpha + \beta + \gamma = p, \ \alpha\beta + \beta\gamma + \gamma\alpha = q, \ \alpha\beta\gamma = r.$$

1. We have

$$(\alpha + \beta + \gamma)^2 = \alpha^2 + \beta^2 + \gamma^2 + 2(\alpha\beta + \beta\gamma + \gamma\alpha)$$

$$\Rightarrow \quad p^2 = \Sigma\alpha^2 + 2q. \text{ Hence } \Sigma\alpha^2 = p^2 - 2q.$$

2. We have

$$(\alpha\beta + \beta\gamma + \gamma\alpha)^2 = \alpha^2\beta^2 + \beta^2\gamma^2 + \gamma^2\alpha^2 + 2(\alpha^2\beta\gamma + \beta^2\gamma\alpha + \gamma^2\alpha\beta)$$

$$\Rightarrow \quad q^2 = \Sigma\alpha^2\beta^2 + 2\alpha\beta\gamma(\alpha + \beta + \gamma)$$

$$\Rightarrow \quad q^2 = \Sigma\alpha^2\beta^2 + 2rp. \text{ Hence } \Sigma\alpha^2\beta^2 = q^2 - 2pr.$$

3. We have

$$(\alpha + \beta)(\beta + \gamma)(\gamma + \alpha) = (p - \gamma)(p - \alpha)(p - \beta). \quad \ldots(1)$$

$$(\because \alpha + \beta + \gamma = p)$$

Since α, β, γ are the roots of $x^3 - px^2 + qx - r = 0$.

$$\therefore \quad x^3 - px^2 + qx - r \equiv (x - \alpha)(x - \beta)(x - \gamma). \quad \ldots(2)$$

Putting $x = p$ in (2), we get

$$(p - \alpha)(p - \beta)(p - \gamma) = p^3 + p^3 + qp - r$$

Hence $(\alpha + \beta)(\beta + \gamma)(\gamma + \alpha) = pq - r$, by (1).

Example 9:

If α, β, γ are the roots of $x^3 + px^2 + qx + r = 0$, find the values of

(1) $\Sigma \dfrac{1}{\alpha}$, (2) $\Sigma\alpha^2$, (3) $\Sigma(\alpha - \beta)^2$, (4) $\Sigma \dfrac{\alpha}{\beta}$, (5) $\Sigma\alpha^2\beta$,

(6) $\Sigma\alpha^3$, (7) $\Sigma \dfrac{\beta^2 + \gamma^2}{\beta\gamma}$ (8) $\Sigma(\beta + \gamma - 2\alpha)$

Solution:

We have

$\alpha + \beta + \gamma = -p, \alpha\beta + \beta\gamma + \gamma\alpha = q, \alpha\beta\gamma = -r.$ ...(1)

1. $\frac{1}{\alpha} = \frac{1}{\alpha} + \frac{1}{\beta} + \frac{1}{\gamma} = \frac{\beta\gamma + \gamma\alpha + \alpha\beta}{\alpha\beta\gamma} = -\frac{q}{r}$, by (1).

2. $(\alpha + \beta + \gamma)^2 = (\alpha^2 + \beta^2 + \gamma^2) + 2(\alpha\beta + \beta\gamma + \gamma\alpha)$

$\Rightarrow$ $(-p)^2 = \Sigma\alpha^2 + 2q$, using (1)

Hence $\Sigma\alpha^2 = p^2 - 2q$.

3. $\Sigma(\alpha + \beta)^2 = (\alpha - \beta)^2 + (\beta - \gamma)^2 + \gamma - \alpha)^2$

$= 2(\alpha^2 + \beta^2 + \gamma) - 2(\alpha\beta + \beta\gamma + \gamma\alpha)$

$= 2(p^2 - 2q) - 2q$, by (1) and (2)

Hence $\Sigma(\alpha + \beta)^2 = 2p^2 -$ **6q**.

4. $\sum\frac{\alpha}{\beta} = \frac{\alpha}{\beta} + \frac{\alpha}{\gamma} + \frac{\beta}{\gamma} + \frac{\beta}{\alpha} + \frac{\gamma}{\alpha} + \frac{\gamma}{\beta}$

$= \frac{\beta + \gamma}{\alpha} + \frac{\gamma + \alpha}{\beta} + \frac{\alpha + \beta}{\gamma}$

$= \frac{-p - \alpha}{\alpha} + \frac{-p - \beta}{\beta} + \frac{-p - \gamma}{\gamma}$ $(\because \alpha + \beta + \gamma = -p)$

$= -p\left(\frac{1}{\alpha} + \frac{1}{\beta} + \frac{1}{\gamma}\right) - 3$

$= -\left(\frac{\beta\gamma + \gamma\alpha + \alpha\beta}{\alpha\beta\gamma}\right) - 3$

$= \frac{pq}{r} - 3$, by (1)

5. $\Sigma\alpha^2\beta = \alpha^2\beta + \alpha^2\gamma + \beta^2\gamma + \beta^2\alpha + \gamma^2\alpha + \gamma^2\beta$

Consider $\Sigma\alpha\Sigma\alpha\beta = (\alpha + \beta + \gamma)(\alpha\beta + \beta\gamma + \gamma\alpha)$

$= \Sigma\alpha_2\beta + 3\alpha\beta\gamma$

$\therefore$ $\Sigma\alpha^2\beta = \Sigma\alpha\Sigma\alpha\beta - 3\alpha\beta\gamma = -pq + 3r$, by (1)

6. Consider $\Sigma\alpha\Sigma\alpha^2 = (\alpha + \beta + \gamma)(\alpha^2 + \beta^2 + \gamma^2)$

$= \Sigma\alpha^3 + \Sigma\alpha^2\beta$

$\therefore$ $\Sigma\alpha_3 = \Sigma\alpha\,\Sigma\alpha^2 - \Sigma\alpha^2\beta$

$= -p(p^2 - 2q) - (-pq + 3r),$

by parts (2) and (v)

Hence $\Sigma\alpha^3 = -p^3 + 3pq - 3r.$

7. $$\sum\frac{\beta^2+\gamma^2}{\beta\gamma} = \frac{\beta^2+\gamma^2}{\beta\gamma} + \frac{\gamma^2+\alpha^2}{\gamma\alpha} + \frac{\alpha^2+\beta^2}{\alpha\beta}$$

$$= \frac{\alpha\left(\beta^2+\gamma^2\right)+\beta\left(\gamma^2+\alpha^2\right)+\gamma\left(\alpha^2+\beta^2\right)}{\alpha\beta\gamma}$$

$$= \frac{\sum\alpha^2\beta}{\alpha\beta\gamma} = \frac{3r-pq}{-r}, \text{ by part (5)}$$

Hence $$\sum\frac{\beta^2+\gamma^2}{\beta\gamma} = \frac{pq-3r}{r}.$$

8. $(\beta + \gamma - 2\alpha)(\gamma + \alpha - 2\beta)(\alpha + \beta - 2\gamma)$

$= (\Sigma\alpha - 3\alpha)(\Sigma\alpha - 3\beta\ (\Sigma\alpha - 3\gamma)$

$= (-p - 3\alpha)(-p - 3\beta)(-p - 3\gamma)$

$= -(p + 3\alpha)(p + 3\beta)(p + 3\gamma)$

$= -[p^3 + 3p^2\,\Sigma\alpha + 9p\Sigma\alpha\beta + 27\alpha\beta\gamma]$

$= -[p^3 - 3p^2.\,p + 9pq - 27r]$, by (1)

$= 2p^3 - - 9pq + 27r.$

Example 10:

If α, β, γ are the roots of $x^3 - px^2 + qx - r = 0$, find the values of (1) $\Sigma\alpha^2\beta^3$, (2) $\Sigma\alpha^2\beta^3$.

Solution:

$\Sigma\alpha^2\beta^2 = q^2 - 2pr.$

Consider $\Sigma\alpha\Sigma\alpha\beta = \Sigma\alpha^2\beta + 3\alpha\beta\gamma$

$\Rightarrow$ $pq = \Sigma\alpha^2\beta + 3r$. Hence $\Sigma\alpha^2\beta = pq - 3r.$

Example 11:

If α, β, γ be the roots of the equation $x^3 + qx + r = 0$, find the values of

1. $\Sigma(\beta + \gamma)^2$;
2. $\Sigma\alpha^2\beta^3$,
3. $\sum\frac{1}{\beta+\gamma}$

4. $\sum\left(\frac{\beta+\gamma}{\gamma+\beta}\right)$

5. $\sum\frac{\beta^2+\gamma^2}{\beta+\gamma}$

6. $\sum\frac{2\beta\gamma-\alpha^2}{\beta+\gamma-\alpha}$.

Solution:

Since α, β, γ are the roots of $x^3 + qx + r = 0$,

$\therefore$ (1) $\alpha+\beta+\gamma = 0$, (2) $\Sigma\alpha\beta + q$, (3) $\alpha\beta\gamma = -r$.

1. From (1), $\beta+\gamma = -\alpha \Rightarrow \Sigma(\beta+\gamma)^2 = \Sigma\alpha^2$

Hence $\Sigma(\beta+\gamma)^2 = (\alpha+\beta+\gamma)^2 - 2\Sigma\alpha\beta = -2q$, by (1) and (2).

2. $\Sigma\alpha^2\beta^2 = (\alpha\beta+\beta\gamma+\gamma\alpha)^2 - 2\alpha\beta\gamma(\alpha+\beta+\gamma) = q^2 - 0 = q^2$.

3. $\sum\frac{1}{\beta+\gamma} = -\sum\frac{1}{\alpha} = -\left(\frac{1}{\alpha}+\frac{1}{\beta}+\frac{1}{\gamma}\right)$

$$= -\left(\frac{\beta\gamma+\gamma\alpha+\alpha\beta}{\alpha\beta\gamma}\right) = \frac{q}{r}, \text{ by (2) and (3).}$$

4. $\sum\left(\frac{\beta}{\gamma}+\frac{\gamma}{\beta}\right) = \sum\frac{\beta^2+\gamma^2}{\beta\gamma}$

$$= \frac{\beta^2+\gamma^2}{\beta\gamma} + \frac{\gamma^2+\alpha^2}{\gamma\alpha} + \frac{\alpha^2+\beta^2}{\alpha\beta}$$

$$= \frac{\alpha\left(\beta^2+\gamma^2\right)+\beta\left(\gamma^2+\alpha^2\right)+\gamma\left(\alpha^2+\beta^2\right)}{\alpha\beta\gamma}$$

$$= \frac{\sum\alpha^2\beta}{-r}. \qquad \ldots(4)$$

Consider $\Sigma\alpha\Sigma\alpha\beta = \Sigma\alpha^2\beta + 3\alpha\beta\gamma$

$\Rightarrow$ $0.q = \Sigma\alpha^2\beta + 3(-r)$, by (1), (2), (3)

$\therefore$ $\Sigma\alpha^2\beta = 3r$.

Hence $\sum\left(\frac{\beta}{\gamma}+\frac{\gamma}{\beta}\right) = -3$, by (4)

5. We have $\beta+\gamma = -\alpha$ and $\beta^2+\gamma^2 = (\beta+\gamma)^2 - 2\beta\gamma = \alpha^2 - 2\beta\gamma$

$$\therefore \quad \sum\frac{\beta^2}{\beta}+\frac{\gamma^2}{\gamma}=\sum\frac{a^2-2\beta\gamma}{-\alpha}=-\sum\alpha+2\sum\frac{\beta\gamma}{\alpha}$$

$$=0+2\left(\frac{\beta^2}{\alpha}+\frac{\gamma\alpha}{\beta}+\frac{\alpha\beta}{\gamma}\right)$$

$$=\frac{2}{\alpha\beta\gamma}\sum\beta^2\gamma^3=\frac{-2q^2}{r}, \text{ by part (2).}$$

6. From (1), $\beta+\gamma=-\alpha$ and so $\beta+\gamma-\alpha=-2\alpha$.

$$\therefore \sum\frac{2\beta\gamma-\alpha^2}{\beta+\gamma-\alpha}=\sum\frac{2\beta\gamma-\alpha^2}{-2\alpha}=\frac{1}{2}\sum\alpha-\sum\frac{\beta\gamma}{\alpha}$$

$$=\frac{1}{2}\times 0-\left(\frac{\beta\gamma}{\alpha}+\frac{\gamma\alpha}{\beta}+\frac{\alpha\beta}{\gamma}\right), \text{ using (1)}$$

$$=-\frac{\beta^2\gamma^2+\gamma^2\alpha^2+\alpha^2\beta^2}{\alpha\beta\gamma}=\frac{q^2}{r}, \text{ by part (2).}$$

Example 12:

If $\alpha, \beta, \gamma, \delta$ be the roots of the equation

$$x^4+px^3+qx^2+rx+s=0,$$

find the values of (1) $\Sigma\alpha^2$, (2) $\Sigma 1/\alpha$, (3) $\Sigma\alpha^2\beta$,

(4) $\Sigma\alpha^2\beta\delta$, (5) $\Sigma\alpha^2\beta^2$, (6) $\Sigma\alpha^4$.

Solution:

We have $\Sigma\alpha=\alpha+\beta+\gamma+\delta=-p$, ...(1)

$\Sigma\alpha\beta=\alpha\beta+\alpha\gamma+\alpha\delta+\beta\gamma+\gamma\delta+\gamma\delta=q$, ...(2)

$\Sigma\alpha\beta\gamma=\alpha\beta\gamma+\alpha\beta\delta+\alpha\gamma\delta+\beta\gamma\delta=-r$, ...(3)

$\alpha\beta\gamma\delta=s$. ...(4)

1. $\Sigma\alpha^2=\alpha^2+\beta^2+\gamma^2+\delta^2$

$$=(\alpha+\beta+\gamma+\delta)^2-2(\alpha\beta+\alpha\gamma+\alpha\delta+\beta\gamma+\beta\delta+\gamma\delta)$$

$$=p^2-2q. \text{ by (1) and (2).}$$

2. $\sum\frac{1}{\alpha}=\frac{1}{\alpha}+\frac{1}{\beta}+\frac{1}{\gamma}+\frac{1}{\delta}$

$$=\frac{\beta\gamma\delta+\alpha\gamma\delta+\alpha\gamma\delta+\alpha\beta\gamma}{\alpha\beta\gamma\delta}=-\frac{r}{s}, \text{ by (3) and (4).}$$

3. Consider the identity

$\Sigma\alpha\ \Sigma\beta = \Sigma\alpha^2\beta + \Sigma\alpha\beta\gamma$

$\therefore \quad \Sigma\alpha^2\beta = \Sigma\alpha\ \Sigma\alpha\beta - \Sigma\alpha\beta\gamma.$

Hence $\Sigma\alpha^2\beta = -pq + r$, by (1), (2), (3).

4. Consider the identity

$\Sigma\alpha\Sigma\alpha\beta\gamma = \Sigma\alpha^2\beta\gamma + 4\alpha\beta\gamma\delta$

$\therefore \quad \Sigma\alpha^2\beta\gamma = \Sigma\alpha\Sigma\alpha\beta\gamma - 4\alpha\beta\gamma\delta$

$= (-p)(-r) - 45$, by (1), (3), (4).

Hence $\Sigma\alpha^2\beta\gamma = pr - 4s.$ ...(5)

5. Consider the identity

$(\Sigma\alpha\beta)^2 = \Sigma\alpha^2\beta^2 + 2\ \Sigma\alpha^2\beta\gamma + 6\alpha\beta\gamma\delta.$

$\therefore \quad \Sigma\alpha^2\beta^2 = (\Sigma\alpha\beta)^2 - 2\Sigma\alpha^2\beta\gamma - 6\alpha\beta\gamma\delta$

$= q^2 - 2(pr - 4s) - 6s$, by (2), (4), (5).

Hence $\Sigma\alpha^2\beta^2 = q^2 - 2pr + 2s.$

6. We have $\Sigma\alpha^2\ \Sigma\alpha^2 = \Sigma\alpha^4 + 2\Sigma\alpha^2\beta^2.$

$\therefore \quad \Sigma\alpha^4 = (p^2 - 2q)^2 - 2(q^2 - 2pr + 2s)$, by parts (1) (5).

Hence $\Sigma\alpha^4 = p^4 + 2p^2 - 4p^2q + 4pr - 4s.$

Example 13:

If α, β, γ be the roots of $x^3 + px^2 + qx + r = 0$; find the value of

$$\sum\frac{\beta^2+\gamma^2}{\beta+\gamma}.$$

Solution:

We have $^3 + px^2 + qx + r = 0$

$\alpha + \beta + \gamma = -p,\ \Sigma\alpha\beta = q,\ \alpha\beta\gamma = -r.$

$$\text{Now} \quad \frac{\beta^2+\gamma^2}{\beta+\gamma} = \frac{(\beta+\gamma)^2 - 2\beta\gamma}{\beta+\gamma} = (\beta+\gamma) - \frac{2\beta\gamma}{\alpha+\beta}.$$

$$\therefore \sum\frac{\beta^2+\gamma^2}{\beta+\gamma} = \sum(\beta+\gamma) - 2\left[\frac{\beta\gamma}{\beta+\gamma} + \frac{\gamma\alpha}{\gamma+\alpha} + \frac{\alpha\beta}{\alpha+\beta}\right]$$

$$= 2\Sigma\alpha - \frac{2\sum\beta\gamma(\gamma+\alpha)(\alpha+\beta)}{(\beta+\gamma)(\gamma+\alpha)(\alpha+\beta)}$$

$$= -2p - \frac{2\sum\beta\gamma\left(\sum\alpha\beta + \alpha^2\right)}{(-p-\alpha)(-p-\beta)(-p-\gamma)}$$

$$= -2p + \frac{2\sum\beta\gamma\left(q + \alpha^2\right)}{(p+\alpha)(p+\beta)(p+\gamma)}$$

$$= -2p + \frac{2\left(q.q + \alpha\beta\gamma\sum\alpha\right)}{p^3 + p^2\sum\alpha + p\sum\alpha\beta + \alpha\beta\gamma}$$

$$= -2p + \frac{2\left(q^2 + pr\right)}{p^3 - p^3 + pq - r}$$

$$= \frac{2p^2 + 4pr - 2p^2q}{pq - r}.$$

Example 14:

If α, β, γ are the roots of $x^3 + qx + r = 0$, prove that

$$\frac{\alpha^7 + \beta^2 + \gamma^7}{7} = \frac{\alpha^2 + \beta^2 + \gamma^2}{2} \cdot \frac{\alpha^5 + \beta^5 + \gamma^5}{5}$$

Solution:

Since α, β, γ are the roots of the equation

$$x^3 + qx + r = 0, \quad \text{...(1)}$$

$\therefore$ $\alpha + \beta + \gamma = 0$, $\Sigma\alpha\beta = q$, $\alpha\beta\gamma = -r$. ...(2)

Now $\alpha^2 + \beta^2 + \gamma^2 = (\alpha + \beta + \gamma)^2 - 2(\alpha\beta + \beta\gamma + \gamma\alpha) = -2q$. ...(3)

Subsisting α, β, γ for x in (1) and adding, we get

$$(\alpha^2 + \beta^2 + \gamma^2) + q(\alpha + \beta + \gamma) + 3r = 0$$

$\Rightarrow$ $\alpha^3 + \beta^3 + \gamma^3 = -3r$ ...(4)

Multiplying (1) throughout by x, we get

$$x^4 + qx^2 + rx = 0. \quad \text{...(5)}$$

Substituting α, β, γ for x in (5) and adding, we get

$$(\alpha^4 + \beta^4 + \gamma^4) + q(\alpha^2 + \beta^2 + \gamma^2) + r(\alpha + \beta + \gamma) = 0.$$

Using (2) and (3), we get $\alpha^4 + \beta^4 + \gamma^4 = 2q^2$. ...(6)

Multiplying (1) throughout by x^2, we get

$$x^5 + qx^3 + rx^2 = 0. \quad \text{...(7)}$$

Substituting α, β, γ for x in (7) and adding, we get

$$(\alpha^5 + \beta^5 + \gamma^5) + q(\alpha^3 + \beta^3 + \gamma^3) + r(\alpha^2 + \beta^2 + \gamma^2) = 0.$$

$$\therefore \quad \alpha^5 + \beta^5 + \gamma^5 = -q(\alpha^3 + \beta^3 + \gamma^3) - r(\alpha^2 + \beta^2 + \gamma^2).$$

Using (3) and (4), $\alpha^5 + \beta^5 + \gamma^5 = 5qr.$ ...(8)

Multiplying (1) throughout by x^4, we obtain

$$x^7 + qx^5 + rx^4 = 0. \qquad ...(9)$$

Substituting α, β, γ for x in (9) and adding, we get

$$\Sigma\alpha^7 + q\Sigma\alpha^5 + r\Sigma\alpha^4 = 0.$$

$$\therefore \quad \Sigma\alpha^7 = -q(5qr) - r(2q^2) = -7q^2 r, \text{ by (8) and (6)}$$

$$\text{Hence } \frac{1}{7}\sum a^7 = -q^2 r = \left(\frac{-2q}{2}\right)\left(\frac{5qr}{5}\right) = \left(\frac{1}{2}\sum\alpha^2\right)\left(\frac{1}{5}\sum\alpha^5\right).$$

Example 15:

Form the cubic whose roots are the values of a, b, g given by the relations $\alpha + \beta + \gamma = 3$, $\alpha^2 + \beta^2 + \gamma^2 = 5$, $\alpha^3 + \beta^3 + \gamma^3 = 11$.

Hence find the value of $\alpha^4 + \beta^4 + \gamma^4$.

Solution:

Suppose the cubic equation whose roots are α, β, γα be

$$x^3 + px^2 + qx + r = 0. \qquad ...(1)$$

Then $\alpha + \beta + \gamma = -p$, $\Sigma\alpha\beta = q$, $\alpha\beta\gamma = -r$. ...(2)

Since $\alpha + \beta + \gamma = 3$, so $p = -3$, by (2).

We have $\alpha^2 + \beta^2 + \gamma^2 = (\alpha + \beta + \gamma)^2 - 2(\alpha\beta + \beta\gamma + \gamma\alpha)$.

$$\Rightarrow \quad 5 = 9 - 2q \quad \Rightarrow \quad q = 2.$$

Since α, β, γ are the roots of (1), therefore by

$$(\alpha^3 + \beta^3 + \gamma^3) + p(\alpha^2 + \beta^2 + \gamma^2) + q(\alpha + \beta + \gamma) + 3r = 0.$$

$$\Rightarrow \quad 11 + (-3)(5) + 2(3) + 3r = 0, \text{ or } r = -2/3.$$

Now from (1), the required cubic equation is

$$x^3 - 3x^2 + 2x - \frac{2}{3} = 0. \qquad ...(3)$$

Multiplying (3) throughout by x, we have

$$x^4 - 3x^3 + 2x^2 - \frac{2}{3}x = 0$$

$$\Rightarrow \quad x^4 = 3x^3 - 2x^2 + \frac{2}{3}x. \qquad ...(4)$$

Substituting α, β, γ for x in (4) and adding,

$$\alpha^4 + \beta^4 + \gamma^4 = 3(\alpha^3 + \beta^3 + \gamma^3) - 2(\alpha^2 + \beta^2 + \gamma^2) + \frac{2}{3}(\alpha + \beta + \gamma)$$

$$= 3(11) - 2(5) + \frac{2}{3}(3) = 25.$$

1.5 SUM OF POWERS OF THE ROOTS OF AN EQUATION

Suppose α, β, γ are the roots of a cubic equation.

Then the values of the expressions such as

$$\alpha^4 + \beta^4 + \gamma^4,\ \alpha^5 + \beta^5 + \gamma^5,\ \alpha^{-4} + \beta^{-4} + \gamma^{-4}$$

etc., can be easily found with the help of the following:

Rule:

1. If $\alpha_1, \alpha_2, ..., \alpha_n$ be the roots of the equation f (x) = 0, and k be any positive integer, then $S_k \equiv \alpha_1^k + \alpha_2^k + ... + \alpha_n^k$ is the coefficient of $\frac{1}{x^{k+1}}$ in the expansion of $\frac{f'(x)}{f(x)}$ in descending powers of x.

2. Further if $\alpha_1, \alpha_2, ..., \alpha_n$ be all non-zero, then

$$S_{-k} \equiv \alpha_1^{-k} + \alpha_2^{-k} + ... + \alpha_n^{-k}$$

is the coefficient of x^{k-1} in the expansion of $-\frac{f'(x)}{f(x)}$ in ascending powers of x.

Example 1:

If α, β, γ are the roots of $x^3 + px^3 + qx + r = 0$, form the equations whose roots are

1. $\frac{\alpha}{\beta + \gamma - \alpha}, \frac{\beta}{\gamma + \alpha - \beta}, \frac{\gamma}{\alpha + \beta - \gamma},$

2. $\alpha^2 + 2\beta\gamma, \beta^2, = 2\gamma\alpha, \gamma^2 + 2\alpha\beta.$

Solution:

We have α + β + γ = – p and αβγ = – r.

1. Let $y = \frac{\alpha}{\beta + \gamma - \alpha} = \frac{\alpha}{(-p - \alpha) - \alpha} = \frac{\alpha}{-p - 2\alpha}$

We may take $y = y = \frac{x}{-p - 2x} \Rightarrow x = -\frac{py}{1 + 2y}.$

Putting this value of x in $x^3 + px^2 + qx + r = 0$, and simplifying,

$$-p^3y^3 + p^3y^2(1 + 2y) - pqy(1 + 2y)^2 + r(1 + 2y)^3 = 0.$$

Hence $(p^3 - 4pq + 8r)y^3 + (p^3 - 4pq + 12r)y^2 + (6r - pq)y + r = 0$ is the required equation.

2. Let $y = \alpha^2 + 2\beta\gamma = \frac{1}{\alpha}(\alpha^2 + 2\alpha\beta\gamma) = \frac{1}{\alpha}(\alpha^3 - 2r)$

$\therefore \alpha^3 - \alpha y - 2r = 0.$ ...(1)

Since a is a root of $x^3 + px^2 + qx + r = 0$, therefore

$\alpha^3 + p\alpha^2 + q\alpha + r = 0$ ...(2)

Subtracting (1) from (2), we obtain

$p\alpha^2 + (q + y)\alpha + 3r = 0.$ ...(3)

Multiply (3) by α, (1) by p and subtract,

$(q + y)\alpha^2 + (3r + py)\alpha + 2pr = 0.$ ...(4)

From (3) and (4), we obtain

$$\frac{\alpha}{2pr(q + y) - 3r(3r + py)} = \frac{a}{3r(q + y) - 2p^2r}$$

$$= \frac{1}{p(3r + py) - (q + y)^2}.$$

$$\Rightarrow \quad \alpha^2 = \frac{2pqr - 9r^2 - pry}{p(3r + py) - (q + y)^2}, \alpha = \frac{3r(q + y) - 2p^2r}{p(3r + py) - (q + y)^2}$$

Eliminating a, we obtain

$(2pqr - 9r^2 - pry)[p(3r + py) - (q + y)^2] = [3r(q + y) - 2p^2r]^2$, which is the required equation.

Example 2:

Transform the equation

$$x^2 - \frac{5}{2}x^2 - \frac{7}{18}x + \frac{1}{108} = 0$$

into one with integral coefficients and the leading coefficient unity.

Solution:

We have $x^2 - \frac{5}{2}x^2 - \frac{7}{18}x + \frac{1}{108} = 0$

Taking $y = kx$ or $x = y/k$ in the given equation,

$$\left(\frac{y}{k}\right)^3 - \frac{5}{2}\left(\frac{y}{k}\right)^2 - \frac{7}{18}\left(\frac{y}{k}\right) + \frac{1}{108} = 0$$

$$\Rightarrow \quad y^2 - \frac{5k}{2}y^2 - \frac{7k^2}{18}y + \frac{1}{108} = 0$$

$$\Rightarrow \quad y^2 - \frac{5ky^2}{2^1.3^0} - \frac{7k^2y}{2^1.3^2} + \frac{k^2}{22.3^3} = 0 \qquad ...(1)$$

The value of k for which the fractions will disappear is k = 2.3 = 6. Putting this value of k in (1), we get

$$y^3 - 15y^2 - 14y + 2 = 0.$$

Hence $x^3 - 15x^2 - 14x + + 2 = 0$ is the required equation.

Example 3:

Find the equation whose roots are the squares of the roots of $x^3 + x^3 + x^2 + 2x + 3 = 0$.

Solution:

We have $x^3 + x^3 + x^2 + 2x + 3 = 0$

Putting $y = x^2$ in the given equation, we get

$$y^2x + xy + y + 2x + 3 = 0 \Rightarrow x\,(y^2 + y + 2) = -\,(y + 3).$$

Squaring both sides, $x^2(y^2 + y + 2)^2 = (y + 3)^2$

$$\Rightarrow \quad y(y^2 + y + 2)^2 = (y + 3)^2.$$

Hence $y^5 + 2y^4 + 5y^3 - 2y - 9 = 0$ is the required equation.

Example 4:

Transform the equation $25x^4 + 5x^3 - 7x^2 + 1 = 0$ into one with integral coefficients and unity for the coefficient of the first term.

Solution:

We have $25x^4 + 5x^3 - 7x^2 + 1 = 0$

The given equation is expressible as

$$x^3 + \frac{1}{5}x^3 - \frac{7}{25}x^2 + \frac{1}{25} = 0. \qquad ...(1)$$

Let y = kx or x = y/k. Putting in (1), we get

$$\left(\frac{y}{k}\right)^4 + \frac{1}{5}\left(\frac{y}{k}\right)^3 - \frac{7}{25}\left(\frac{y}{k}\right)^2 + \frac{1}{25} = 0$$

$$y^4 + \frac{1}{5}ky^3 - \frac{7}{25}k^2y^2 + \frac{k^2}{25} = 0. \qquad ...(2)$$

Here the coefficient of the first term is unity.

If we take k = 5 in (2), we obtain

$$y^4 + y^3 - 7y^2 + 25 = 0.$$

Hence $x^4 + x^3 - 7x^2 + 25 = 0$

is the required equation with integral coefficients and unity for the coefficient of the first term.

Example 5:

Find the equation whose roots are the cubes of the roots of the equation $x^4 - x^3 + 2x^2 + 3x + 1 = 0$.

Solution:

We have $x^4 - x^3 + 2x^2 + 3x + 1 = 0$

Taking $y = x^3$ in the given equation, we get

$$yx - y + 2x^2 + 3x + 1 = 0$$

$$\Rightarrow \quad x\,[(y + 3) + 2x] = y - 1. \qquad ...(1)$$

Cubing both sides, we obtain

$$x^3\ [(y + 3)^3 + 8x^3 + 3\ (y + 3).\ 2x\ \{(y + 3) + 2x\}] = (y - 1)^3$$

$$\Rightarrow \quad y\ \{(y + 3)^3 + 8y + 6\ (y + 3)\ (y - 1)] = (y - 1)^3, \text{ using (1)}$$

$$\Rightarrow \quad y\ [(x^3 + 9x^2 + 27y + 27) + 6y^2 + 20y - 18] = y^3 - 3y^3 + 3y - 1.$$

Hence $y^4 + 14y^3 + 50y^2 + 6y + 1 = 0$ is the required equation.

Example 5:

If α, β, γ are the roots of $x^3 + px^2 + qx + r = 0$, *find the equation whose roots are*

$$\alpha - \frac{1}{\beta\gamma}, \beta - \frac{1}{\gamma\alpha}, \gamma - \frac{1}{\alpha\beta}$$

Solution:

$$y = a - \alpha - \frac{1}{\beta\gamma} = \alpha - \frac{1}{\alpha\beta\gamma} = \alpha\left(1 + \frac{1}{r}\right) \qquad [\because \alpha\beta\gamma = -r]$$

We may take

$$y = x\left(1 + \frac{1}{r}\right) \text{ or } x = \frac{ry}{1 + r}.$$

Putting this value of x in the given equation, we get

$$\frac{r^2y^2}{(1+r)^3} + p\frac{r^2y^2}{(1+r)^2} + q\frac{ry}{1+r} + r = 0.$$

Hence $r^2x^3 = pr(1 + r)x^2 + q(1 + r)^2 x + (1 + r)^3 = 0.$

Example 6:

If α, β, γ, be the roots $x^3 - px^2 + qx - r = 0$,

(a) find the equation whose roots are $\alpha^2, \beta^2, \gamma^2$

(b) find the equation whose roots are

$\beta\gamma + 1/\alpha, \gamma\alpha + 1/\beta, \alpha\beta + 1/\gamma.$

Solution:

(a) In other words, we have to find the equation whose roots are the squares of the roots of

$$x^3 - px^2 + qx - r = 0. \qquad ...(1)$$

Taking $y = x^3$ in (1), we get

$xy - py + qx - r = 0$

$\Rightarrow \quad (y + q)x = py + r$

$\Rightarrow \quad (y + q)^2 x^2 = (py + r)^2$. Hence the required equation is

$y(y + q)^2 - (py + r)^2 = 0.$

(b) $y = \beta\gamma + \dfrac{1}{\alpha} = \dfrac{\alpha\beta\gamma + 1}{\alpha} = \dfrac{r+1}{\alpha}.$

Thus, we may take $y = y = \dfrac{r+1}{x} \Rightarrow x = \dfrac{r+1}{y}.$

Substituting this value of x in (1), we get

$$\left(\frac{r+1}{y}\right)^3 - p\left(\frac{r+1}{y}\right)^2 + q\left(\frac{r+1}{y}\right) - r = 0.$$

Hence $ry^3 - q(r + 1)y^2 + p(r + q)^2 y - (r + 1)^2 = 0$

is the required equation.

Example 7:

Find the equation whose roots are the squares of the roots of the equation $x^3 - x^2 - 8x + 6 = 0$.

Solution:

We have $x^3 - x^2 - 8x + 6 = 0$

Let $y = x^2$. We may write the given equation as

$$y.x - y - 8x + 6 = 0$$

$\Rightarrow x(y - 8) = y - 6.$

Squaring both sides, we obtain

$$x^2(y - 8)^2 = (y - 6)^2$$

$\Rightarrow y(y - 8)^2 = (y - 6)^2.$

Hence $y^3 - 17y^2 + 76y - 36 = 0$ is the required equation.

Example 8:

Find an equation whose roots are the roots of $4x^5 - 2x^3 + 7x - 3 = 0$ each increased by 2.

Solution:

Here $y = x + 2$ so that $x = y - 2$. Putting this value of x in $4x^5 - 2x^3 + 7x - 3 = 0$, we get

$$4(y - 2)^5 - 2(y - 2)^3 + 7(y - 2) - 3 = 0$$

$\Rightarrow \quad 2(y - 2)^2[2(y - 2)^3 - (y - 2)] + 7y - 17 = 0$

$\Rightarrow \quad (y^2 - 8y + 8)(2y^3 - 12y^2 + 23y - 14) + 7y - 17 = 0$

$\Rightarrow \quad 4y^5 - 40y^4 + 158y^3 - 308y^2 + 303y - 129 = 0.$

Interchanging y and x, the required equation is

$$4x^5 - 40x^4 + 158x^3 + 308x^2 + 303x - 129 = 0.$$

Example 9:

Form an equation whose roots are the reciprocals of the roots of the equation $x^3 - x^2 + 2x + 1 = 0$.

Solution:

We have $x^3 - x^2 + 2x + 1 = 0$

If y be a root of the required equation, then $y = 1/x$.

Putting $x = 1/y$ in the given equation, we obtain

$$\left(\frac{1}{y}\right)^2 - \left(\frac{1}{y}\right)^2 + 2\left(\frac{1}{y}\right) + 1 = 0.$$

Hence $y^3 + 2y^2 - y + 1 = 0$ is the required equation.

Example 10:

Form an equation whose roots are roots are three times those of the equation $x^3 + 2x^2 - x + 3 = 0$.

Solution:

If y be a root of the required equation, then $y = 3x$. Putting $x = y/3$ in the given equation, we obtain

$$\left(\frac{y}{3}\right)^3 + 2\left(\frac{y}{3}\right)^2 - \left(\frac{y}{3}\right) + 3 = 0.$$

Hence $y^3 + 6y^2 - 9y + 81 = 0$ is the required equation.

Example 11:

If α, β, γ be the roots of $x^3 + px^2 + qx + r = 0$, form the equation whose roots are $\beta\gamma - \alpha^2$, $\gamma\alpha - \beta^2$, $\alpha\beta - \gamma^2$.

Solution:

$$\text{Let } y\ \beta\gamma - \alpha^2 = \frac{1}{\alpha}\left(\alpha\beta\gamma - \alpha^3\right) = \frac{1}{\alpha}\ (-r - \alpha^2)$$

$\therefore \quad \alpha^3 + \alpha y + r = 0.$ ...(1)

Since α is a roots of $x^2 + px^2 + qx + r = 0$,

$\therefore \quad \alpha^3 + p\alpha^2 + q\alpha + r = 0.$ ...(2)

Subtracting (1) from (2), we obtain

$pa^2 + (q - y)\ \alpha = 0$

$\Rightarrow \quad \alpha\ [pq + (q = y)] = 0$

$\Rightarrow \quad \alpha = (y - q)/p$. Putting this value of a in (2), we get

$$\frac{(y-q)^3}{p^3} + p\,\frac{(y-q)^2}{p^2} + \frac{q\,(y-q)}{p} + r = 0.$$

Hence $(y - q)^3 + p^2\,(y - q)^2 + pq\,(y - q) + rp^2 = 0$.

Example 12:

If α, β, γ be the roots of $x^3 - 3x^2 + 6x - 2 = 0$, form the equation whose roots are $\beta^2 + \gamma^2$, $\gamma^2 + \alpha^2$, $\alpha^2 + \beta^2$.

Solution:

We have $\alpha + \gamma + \gamma = 3$ and $\alpha\beta\gamma = 2$.

Let $y = \beta^2 + \gamma^2 = (\beta + \gamma)^2 - 2\beta\gamma = (3 - \alpha)^2 - 4/\alpha$

$\Rightarrow \quad \alpha y = \alpha (3 - \alpha)^2 - 4$

$\Rightarrow \quad \alpha^3 - 6x^2 + \alpha (9 - y) - 4 = 0.$ ...(1)

Since α is a root of the given equation, therefore

$\alpha^3 - 3\alpha^2 + 6\alpha - 2 = 0.$ ...(2)

Subtracting (1) from (2), $3\alpha^2 + \alpha (y - 3) + 2 = 0$ ...(3)

$\Rightarrow \quad 3\alpha^3 + \alpha^2 (y - 3) + 2\alpha = 0.$ ...(4)

Multiply (1) by 3 and subtract it from (4),

$\alpha^2(y + 6) - 16\alpha + 6 = 0.$ ...(5)

From (3) and (5), we get

$$\frac{\alpha^3}{6y - 18 + 32} = \frac{\alpha}{2y + 12 - 18} = \frac{1}{-48 - (y + 6)(y - 3)}$$

$$\Rightarrow \quad \frac{\alpha^2}{6y + 14} = \frac{\alpha}{2y - 6} = \frac{1}{-y^2 - 3y - 30}$$

$$\Rightarrow \quad \alpha^3 = \frac{6y + 14}{-y^2 - 3y - 30}, \alpha = \frac{2y - 6}{-y^2 - 3y - 30}$$

$(6y + 14)(-y^2 - 3y - 30) = (2y - 6)^2$

Hence $y^3 + 6y^2 + 33y + 76 = 0$ is the required equation.

Example 13:

If α, β, γ be the roots of the equation

$$x^3 - 6x^2 + 11x - 6 = 0, s$$

find the equation whose roots are $\beta^2 + \gamma^2, \gamma^2 + \alpha^2 + \beta^2$.

Solution:

We have $\alpha + \beta + \gamma = 6 \ \alpha\beta\gamma = 6.$ ...(1)

Let $\quad y = \beta^2 + \gamma^2 = (\beta + \gamma)^2 - 2\beta\gamma = (6 - \alpha)^2 - 12/\alpha$, using (1)

$\therefore \quad \alpha y + \alpha (6 - \alpha)^2 - 12$

$\Rightarrow \quad \alpha^3 - 12\alpha^2 + (36 - y)\alpha - 12 = 0.$ ...(2)

Since a is a root of $x^3 - 6x^2 + 11x - 6 = 0$,

$\therefore \quad \alpha^3 - 6\alpha^2 + 11\alpha - 6 = 0.$ ...(3)

Subtracting (2) from (3), we obtain

$6\alpha^2 + (y - 25)\alpha + 6 = 0.$ ...(4)

Multiply (4) by a, (3) by 6 and subtract,

$$(y + 11)\alpha^2 - 60\alpha + 36 = 0. \quad ...(5)$$

Solving (4) and (5), we obtain

$$\frac{\alpha^2}{36(y-25)+360} + \frac{\alpha}{6(y+11)+216} = \frac{1}{-360-(y+11)(y-25)}$$

$$\Rightarrow \quad \frac{\alpha^2}{36(y-15)} = \frac{\alpha}{6(y+25)} = \frac{1}{-y^2+14y-84}.$$

$$\therefore \quad \alpha^2 = \frac{36(y-15)}{-y^2+14y-85}, \alpha = \frac{6(y-25)}{-y^2+14y-85}$$

These equations yield

$$\frac{36(y-15)}{-y^2+14y-85} = \frac{36(y-25)^2}{(-y^2+14y-85)^2}$$

$$\Rightarrow \quad (y-25)^2 = (y-15)(-y^2+14y-85)$$

$$\Rightarrow \quad y^2 - 50y + 625 = -y^3 + 29y^2 - 295y + 1275.$$

Interchanging x and y, the required equation is

$$x^3 - 28x^2 + 245x - 650 = 0.$$

Example 14:

Solve $x^3 + 6x^3 + 12x - 19 = 0$, by removing its second term.

Solution:

Replace x by x + h, the transformed equation is

$$(x+h)^3 + 6(x+h)^2 + 12(x+h) - 19 = 0$$

$$\Rightarrow \quad x^3 + x^2(3h+6) + x(3h^2+12h+12) + (h^3+6h^2+12h-19) = 0 \quad ....(1)$$

To remove the second the second term, we take $3h + 6 = 0$ *i.e.*, $h = 2$.

Putting $h = -2$ in (1). we get

$$x^3 - 27 = 0 \Rightarrow (x-3)(cx^2 + 3x + 0) = 0.$$

Solving $x^2 + 3x + 9 = 0$; $x = \dfrac{-3 \pm \sqrt{9-36}}{2} = \dfrac{-3 \pm 3\sqrt{3i}}{2}$

$$\Rightarrow \quad x = 2\omega, 3\omega^2; \text{ where } \omega = \frac{1}{2}(-1 + i\sqrt{3}). \text{ Also } x = 3.$$

Hence the required roots (given by $y = x - 2$) are

$1, 3\omega - 2$ and $3\omega^2 = 2.$

Example 15:

Solve $x^4 - 12x^3 + 49x^2 - 78x + 40 = 0$, by removing its second term.

Solution:

The sum of four roots of the given equation is 12. If we decrease the roots of the given equation by 3, the second term of the required equation would be removed.

Let $y = x - 3$ so that $x = y + 3$. Putting the value of x in the given equation, we get

$$(y + 3)^4 - 12 (y + 3)^3 + 49 (y + 3)^2 - 78 (y + 3) + 40 = 0$$

$$\Rightarrow \quad y^4 - 5y^2 + 4 = 0$$

$$\Rightarrow \quad (y^2 - 4) (y^2 - 1) = 0$$

$$\Rightarrow \quad y = \pm 2, \pm 1.$$

Hence $x = y + 3$ gives 5, 1, 4 and 2 as the required roots of the given equation.

Example 16:

Remove the second term from the equation

$$x^3 - 6x^2 + 4x - 7 = 0.$$

Solution:

Replace x by x + h, the transformed equation is

$$(x + h)^3 - 6(x + h)^2 + 4(x + h) + 7 = 0$$

$$\Rightarrow \quad x^3 + x^2(3h - 6) + x(3h^2 - 12h + 4) + (3h^3 - 6h^2 + 4h - 7) = 0 \text{....(1)}$$

The second term will be missing if $3h - 6 = 0$ *i.e.*, $h = 2$.

Putting $h = 2$ in (1), the required equation is

$$x^3 - 8x - 15 = 0.$$

1.6 COMMON ROOTS AND REPEATED ROOTS OF POLYNOMIAL EQUATIONS

Rule 1: *The common roots of two given polynomials are the roots of their greatest common divisor.*

Rule 2: *To find the repeated roots of a polynomial equation $f(x) = 0$, we first find the g.c.d. $h(x)$ and $f'(x)$. If α be a root of $h(x) = 0$ repeated n times, then α is a roots of $f(x) = 0$ repeated $(n + 1)$ times.*

Example 1:

Solve the equation $16x^4 - 24x^2 + 16x - 3 = 0$, by testing for its repeated roots.

Solution:

Let $f(x) = 16x^4 - 24x^2 + 16x - 3$

$\therefore f'(x) = 64x^3 - 48x + 16 = 16(4x^3 - 3x + 1)$

Now we shall find the g.c.d. of f (x) and f' (x).

$$\begin{array}{r|l} & \quad 4x \\ \hline 2x^3 - 3x + 1 & 16x^4 - 24x^2 + 16x - 3 \\ & 16x^4 - 12x^2 + 4x \\ & \;\;- \quad\; + \quad\; - \\ \hline & -12x^2 + 12x - 3 = -3(4x^3 - 4x + 1) \end{array}$$

$$\begin{array}{r|l} & \quad x + 1 \\ \hline 4x^2 - 4x + 1 & 4x^3 - 3x + 1 \\ & 4x^3 - 4x + x \\ & \;- \quad + \quad - \\ \hline & 4x^2 - 4x + 1 \\ & 4x^2 - 4x + 1 \\ \hline & \qquad 0 \\ \hline \end{array}$$

The g.c.d of f (x) f' (x) is $4x^2 - 4x + 1 = (2x - 1)^2$.

It follows that $x = \frac{1}{2}$ is the only common roots. Since $x = \frac{1}{2}$ is a common root of f (x) = 0 and f' (x) = 0 (occurring twice), so $\frac{1}{2}$ a root of f (x) = 0 repeated three times. (**Rule 2**). If the fourth root of f (x) = 0 is α. Then

$$\text{sum of the roots } = \frac{1}{2} + \frac{1}{2} + \frac{1}{2} + \alpha = 0 \text{ } i.e. \text{ } \alpha = -\frac{3}{2}.$$

$$\text{Hence the roots of the given equation are } = \frac{1}{2}, \frac{1}{2}, \frac{1}{2}, -\frac{3}{2}.$$

Example 2:

The equations

$$6x^3 - 17x^2 + 11x - 2 = 0, \; 12x^3 - 4x^2 - 3x + 1 = 0$$

have two common roots. Determine them and hence solve them.

Solution:

$$\begin{array}{r|l} & \quad 2 \\ 6x^3 - 17x^2 + 11x - 2 & 12x^3 - 4x^2 - 3x + 1 \\ & 12x^3 - 34x^2 + 22x - 4 \\ & - \quad + \quad - \quad + \\ \hline & 30x^2 - 25x + 5 - 5 \; (6x^2 - 5x + 1) \end{array}$$

$$\begin{array}{r|l} & x - 2 \\ 6x^2 - 5x + 1 & 6x^3 - 17x^2 + 11x - 2 \\ & 6x^3 - 5x^2 + x \\ & - \quad + \quad - \\ \hline & -12x^2 + 10x - 2 \\ & -12x^2 + 10x - 2 \\ & + \quad - \quad + \\ \hline & 0 \\ \hline \end{array}$$

The g.c.d. of the given polynomial equation is

$$6x^2 - 5x + 1 = (3x - 1)(2x - 1).$$

Hence the common roots of the two equations are $\frac{1}{2}, \frac{1}{3}$. (Rule 1)

We can write $6x^3 - 17x^2 + 11x - 2 = (6x^2 - 5x + 1)(x - 2)$.

Hence the roots of $6x^3 - 17x^2 + 11x - 2 = 0$ are $\frac{1}{2}, \frac{1}{3}, 2$.

Also $12x^3 - 4x^2 - 3x + 1 = (6x^2 - 5x + 1)(2x + 1)$

Hence the roots of $12x^3 - 4x^2 - 3x + 1 = 0$ are $\frac{1}{2}, \frac{1}{3}, -\frac{1}{2}$.

Example 3:

Solve the equation $4x^3 - 12x^2 - 15x - 4 = 0$, by testing for its repeated roots.

Solution:

Let $f(x) = 4x^3 - 12x^2 - 15x - 4$

$\therefore \quad f'(x) = 12x^2 - 24x - 15 = 3\,(4x^2 - 8x - 5)$

Now we shall find the g.c.d. of f (x) and f' (x).

$$\begin{array}{r|l} & x - 1 \\ \hline 4x^2 - 8x - 5 & 4x^3 - 12x^2 - 15x - 4 \\ & 4x^3 - 8x^2 - 5x \\ & -\quad +\quad + \\ \hline & -4x^2 - 10x - 4 \\ & -4x^2 - 8x - 5 \\ & -\quad +\quad - \\ \hline & -18x - 9 \\ & = -9\,(2x - 1) \end{array} \qquad \begin{array}{r|l} & 2x - 5 \\ \hline 2x + 1 & 4x^2 - 8x - 5 \\ & 4x^2 + 2x \\ & -\quad - \\ \hline & -10x - 5 \\ & -10x - 5 \\ & +\quad + \\ \hline & 0 \\ \hline \end{array}$$

We see that 2x + 1 is the g.c.d of f (x) and f '(x).

Consequently, $x = -\frac{1}{2}$ is the only common roots. Since $x = -\frac{1}{2}$ is a common root of f (x) = 0 and f ' (x) = 0 (occurring only once), so $-\frac{1}{2}$ is a roots of f (x) = 0 repeated twice (**Rule 2**). If the third root of f(x) = 0 is α, then

$$\text{sum of the roots} - -\frac{1}{2} - \frac{1}{2} + \alpha = \frac{12}{4} = 3 \text{ i.e., } \alpha = 4$$

Hence the roots of the given equation are $-\frac{1}{2}, -\frac{1}{2}, 4$.

SOME SOLVED EXAMPLES

Example 1:

Find a necessary condition for the roots of the equation $x^4 + px^3 + qx^2 + rx + s = 0$ to be Arithmetic progression.

Solution:

Let the given roots be $\alpha = a - d$, $\gamma = a$, $\delta = a + d$, $\beta = a + 2d$. Then

$$\alpha + \beta + \gamma + \delta. \qquad ...(1)$$

We have $\quad \alpha + \beta + \gamma + \delta = -p \qquad ...(2)$

$$(\alpha + \beta)(\gamma + \delta) + \alpha\beta + \gamma\delta = q \qquad ...(3)$$

$$\alpha\beta(\gamma + \delta) + \gamma\delta(\alpha + \beta) = -r \qquad ...(4)$$

$$\alpha\beta\gamma\delta = s. \qquad ...(5)$$

From (1) and (2), $\alpha + \beta = \gamma + \delta = -p/2$. ...(6)

From (3) and (6), $(p^2/4) + \alpha\beta = \gamma\delta = q$

$\Rightarrow \quad \alpha\beta + \gamma\delta = q - (p^2/4)$. ...(7)

From (4) and (6), $\alpha\beta + \gamma\delta = 2r/p$. ...(8)

From (7) and (8), $q - \dfrac{p^2}{4} = \dfrac{2r}{p} \Rightarrow 2pq - p^3 = 8r$.

Hence $p^3 - 4pq + 8r = 0$ is the required condition.

Example 2:

Find the equation of lowest degree with rational coefficients having $\sqrt{2} + \sqrt{-3}$ as one if its roots.

Solution:

The remaining roots of the equation of lowest degree having $\sqrt{2} + \sqrt{3}$ i as one of its roots are $\sqrt{2} - \sqrt{3}$ i, $-\sqrt{2} + \sqrt{3}$ i, $-\sqrt{2} - \sqrt{3}$ i.

Hence the required equation is

$$\left[\left(x - \sqrt{2}\right) + \sqrt{3}\,i\right]\left[\left(x - \sqrt{2}\right) - \sqrt{3}\,i\right]\left[\left(x + \sqrt{2}\right) + \sqrt{3}\,i\right]\left[\left(x + \sqrt{2}\right) - \sqrt{3}\,i\right] = 0$$

$$\Rightarrow \quad \left[\left(x - \sqrt{2}\right)^2 + 3\right]\left[\left(x + \sqrt{2}\right)^2 + 3\right] = 0$$

$$\Rightarrow \quad (x^2 - 2)^2 + 3\left(x + \sqrt{2}\right)^2 + 3\left(x - \sqrt{2}\right)^2 + 9 = 0$$

$$\Rightarrow \quad x^4 - 4x^3 + 4 + 6x^2 + 12 + 9 = 0$$

$$\Rightarrow \quad x^4 + 2x^2 + 25 = 0.$$

Example 3:

Find the equation of lowest degree with rational coefficients one root of which is $\sqrt{3} - \sqrt{5}$.

Solution:

The four roots of the required equation are

$$\sqrt{3} - \sqrt{5},\ \sqrt{3} + \sqrt{5},\ -\sqrt{3} + \sqrt{5},\ -\sqrt{3} - \sqrt{5}.$$

Hence the required equation is

$$\left[\left(x-\sqrt{3}\right)+\sqrt{5}\right]\left[\left(x-\sqrt{3}\right)-\sqrt{5}\right]\left[\left(x+\sqrt{3}\right)+\sqrt{5}\right]\left[\left(x+\sqrt{3}\right)-\sqrt{5}\right]=0$$

$$\Rightarrow \quad \left[\left(x-\sqrt{3}\right)^2-5\left(x+\sqrt{3}\right)^2-5\right]=0$$

$$\Rightarrow \quad (x^2-3)^2-5\left(x+\sqrt{3}\right)^2-5\left(x-\sqrt{3}\right)^2+25=0$$

$$\Rightarrow \quad x^4-6x^2+9-10x^2-30+25=0$$

$$\Rightarrow \quad x^4-16x^2+4=0.$$

Example 4:

One root of the equation $x^4 + 2x^3 - 16x^2 - 22x + 7 = 0$ is $2 + \sqrt{3}$. find the other roots.

Solution:

Two roots of the given equation are $2 + \sqrt{3}$, $2 - \sqrt{3}$.

Consequently, $[(x - 2) + \sqrt{3}]\,[(x - 2) - \sqrt{3}] = (x - 2)^2 - 3 = x^2 - 4x + 1$ is a factor of $x^4 + 2x^3 - 16x^2 - 22x + 7$, which on division by $x^2 - 4x + 1$ can be written as

$$x^4 + 2x^3 - 16x^2 - 22x + 7 = (x^2 - 4x + 1)\,(x^2 + 6x + 7).$$

Now $x^2 + 6x + 7 = 0 \Rightarrow x = \dfrac{-6 \pm \sqrt{36-28}}{2} = -3 \pm \sqrt{2}$.

Hence, the other roots are $2 - \sqrt{3}$, $-3 + \sqrt{2}$, $-3 - \sqrt{2}$.

Example 5:

The distances of three points A, B, C on a right line from a fixed origin O on the line are the roots of the equation $ax^3 + 3bx^2 + 3cx + d = 0$. Find the condition that the four points O, A, B, C should form a harmonic division.

Solution:

Let the three points A, B, C be on the x-axis and that their distances from O be α, β, γ, respectively. It is given that α, β, γ are the roots of

$$ax^3 + 3bx^2 + 3cx + d = 0 \qquad \text{...(1)}$$

$$\therefore \quad \alpha\beta + \beta\gamma + \gamma\alpha = \frac{3c}{a},\ \alpha\beta\gamma = -\frac{d}{a}.$$

Since the points O, A, B, C form a harmonic division, so

$$\beta = \frac{2\alpha\gamma}{\alpha + \gamma} \Rightarrow \alpha\beta + \beta\gamma + \gamma\alpha = 3\alpha\gamma \Rightarrow \frac{3c}{a} = 3\alpha\gamma$$

$$\Rightarrow \quad \left(\frac{c}{a}\right) \beta = \alpha\beta\gamma \Rightarrow \frac{c\beta}{a} = -\frac{d}{a} \Rightarrow \beta = -\frac{d}{c}.$$

Since β is a root of (1), $a\beta^3 + 3b\beta^3 + 3c\beta + d = 0$

$\Rightarrow \quad a(-d/c)^3 + 3b(-d/c)^2 + 3c(-d/c) + d = 0.$

Hence $ad^2 - 3bcd + 2c^3 = 0$ is the required condition.

Example 6:

Form an equation whose roots are of the type α/β, β/γ, γ/α; where α, β, γ are the roots of the equation $x^3 + qx + r = 0$.

Solution:

Let $y = \alpha/\beta$ so that $\alpha = \beta y$.

Since α. β are the roots of $x^3 + qx + r = 0$. so

$$\alpha^3 + q\alpha + r = 0, \quad \text{...(1)}$$

$$\beta^3 + q\beta + r = 0. \quad \text{...(2)}$$

Putting $\alpha = \beta y$ in (1) gives $\beta^3\beta^3 + q\beta y + r = 0$. ...(3)

Subtracting (2) from (3), we get

$(y^3 - 1)\beta^3 + q\beta(y - 1) = 0 \Rightarrow \beta(y - 1)[q + (y^2 + y + 1)\beta^2] = 0$

$\Rightarrow \quad q + (y^2 + y + 1)\beta^2 = 0 \; (\because q \neq 0.\ y \neq 1)$

$$\Rightarrow \quad \beta^2 = \frac{-q}{y^2 + y + 1} \quad \text{...(4)}$$

From (2), $\beta(\beta^2 + q) = -r \Rightarrow \beta^2(\beta^2 + q)^2 = r^2$

$$\Rightarrow \quad \left(\frac{-q}{y^2 + y + 1}\right)\left(\frac{-q}{y^2 + y + 1} + q\right)^2 = r^2, \text{ using (4)}$$

$\Rightarrow \quad -q^3(y^2 + y)^2 = r^2(y^2 + y + 1)^3.$

Hence $r^2(y^2 + y + 1)^3 + q^3y^2(y + 1)^2 = 0$ is the required equation.

Example 7:

If α, β, γ are the roots of $x^3 - ax^2 + bx - c = 0$, find the value of $\sum \frac{\alpha^2}{\beta\gamma}$.

Solution:

$$\sum \frac{\alpha^2}{\beta\gamma} = \frac{\alpha^2}{\beta\gamma} + \frac{\beta^2}{\gamma\alpha} + \frac{\gamma}{\alpha\beta} = \frac{\alpha^3 + \beta^3 + \gamma^3}{\alpha\beta\gamma} = \frac{1}{c} \sum \alpha^3 \quad ...(1)$$

We have $\alpha + \beta + \gamma = \alpha$, $\alpha\beta + \beta\gamma + \gamma\alpha = b$ and $\alpha\beta\gamma = c$.

Now $(\alpha + \beta + \gamma)^3 = \sum\alpha^2 + 3 \sum\alpha^2\beta + 6\alpha\beta\gamma$

$\Rightarrow \quad \alpha^3 = \sum\alpha^3 + 3 [\sum\alpha \sum\alpha\beta - 3 \alpha\beta\gamma] + 6\alpha\beta\gamma$

$\Rightarrow \quad \alpha^3 = \sum\alpha^3 + 3 (\alpha\beta - 3c) + 6c.$

$\therefore \sum\alpha^3 = \alpha^3 - 3\alpha\beta + 3c.$ ...(2)

From (1) and (2), $\sum \frac{\alpha^2}{\beta\gamma} = \frac{a^2 - 3ab + 3c}{c}.$

Example 8:

The roots of $16x^4 - 64x^3 + 56x^2 + 16x - 15 = 0$ are A.P. Find them.

(D.U., B.Sc. (G) 1996)

Solution:

Let the required roots be a – d, a, a + d, a 2d. Then

$\sigma_1 \equiv a - d + a + a + d + a + 2d = 64/16$

$\Rightarrow 4a + 2d = 4 \Rightarrow d = 2 - 2a.$ ...(1)

$\sigma_1 \equiv (a + a - d)(a + d + a + 2d)$

$+ a(a - d) + (a + d)(a + 2d) = 56/16$ ...(2)

From (1) and (2), we get

$(2a - 2 + 2a)(2a + 6 - 6a) + a(a - 2 + 2a)$

$+ (a + 2 - 2a)(a + 4 - 4a) = 7/2$

$\Rightarrow \quad (4a - 2)(6 - 4a) + a(3a - 2) + (2 - a)(4 - 3a) = 7/2$

$\Rightarrow \quad -16a^2 + 32a - 12 + 3a^2 - 2a + 3a^2 - 10a + 8 = 7/2$

$\Rightarrow \quad -10a^2 + 20a - 4 = 7/2$

$\Rightarrow \quad 4a^2 - 8a + 3 = 0.$

$\therefore \quad a = \frac{3}{2}, \frac{1}{2}.$

Taking a = 1/2 in (1), we get d = 1. Hence the roots are

$$\frac{1}{2} - 1, \frac{1}{2}, \frac{1}{2} + 1, \frac{1}{2} 2 \Rightarrow -\frac{1}{2}, \frac{1}{2}, \frac{3}{2}, \frac{5}{2}.$$

Example 9:

If α, β, γ are the roots of $x^3 + 5x^2 - 6x + 3 = 0$, find $\Sigma\alpha^{-3}$.

Solution:

Let $f(x) = x^3 + 5x^2 - 6x + 3$ so that $f'(x) = 3x^2 + 10x - 6$.

Now $$-\frac{f'(x)}{f(x)} = \frac{6 - 10x = 3x^2}{3 - 6x + 5x^2 + x^3}$$

$$= \frac{6}{3}\left(1 - \frac{10x + 3x^2}{6}\right)\left[1 - \left(\frac{6x - 5x^2 - x^3}{3}\right)\right]^{-1}$$

$$= \frac{6}{3}\left(1 - \frac{10x + 3x^2}{6}\right)\left[1 + \left(\frac{6x - 5x^2 - x^3}{3}\right) + \left(\frac{6x - 5x^2 - x^3}{3}\right) + \ldots\right]$$

...(1)

By the Rule (2), $\alpha^{-3} + \beta^{-3} + \gamma^{-3}$ is the coefficient of x^2 in the expansion of $-f'(x)/f(x)$ in ascending powers of x.

$$\therefore \alpha^{-3} + \beta^{-3} + \gamma^{-3} = 2\left[\left\{1\left(-\frac{5}{3}\right) + 1\left(\frac{6}{3}\right)^2\right\} - \frac{10}{6}\left(\frac{6}{3}\right) - \frac{3}{6}.1\right]$$

$$= 2\left(-\frac{5}{3} + 4 - \frac{10}{3} - \frac{1}{2}\right) = 2\left(-5 + \frac{7}{2}\right) = -3.$$

Example 10:

Solve the equation: $3x^4 - 40x^3 + 130x^3 - 120x + 27 = 0$, given that the product of two of its roots is equal to the product of the other two.

(D.U., B.Sc. (H) 1996)

Solution:

Let required roots be α, β, γ, δ. It is given that

$\alpha\beta = \gamma\delta.$ ...(1)

Now $\sigma_1 \equiv \alpha + \beta + \gamma + \delta = 40/3$ ...(2)

$\sigma_2 \equiv (\alpha + \beta)(\gamma + \delta) + \alpha\beta + \gamma\delta = 130/3$ ...(3)

$\sigma_3 \equiv (\alpha + \beta)\gamma\delta + (\gamma + \delta)\alpha\beta = 40$ ...(4)

$\sigma_4 \equiv \alpha\beta\gamma\delta = 9.$ ...(5)

From (1) and (5), $\alpha\beta = \gamma\delta = 3.$ ...(6)

From (3) and (6), $(\alpha + \beta)(\gamma + \delta) = \frac{130}{3} - 6 = \frac{112}{3}.$ ...(7)

From (2) and (7), it follows that $(\alpha + \beta)$ and $(\gamma + \delta)$ are the roots of

$3y^2 - 40y + 112 = 0$

$\Rightarrow$ $(3y - 28)(y - 4) = 0$

$\Rightarrow$ $y = 28/3, 4.$

We take $\alpha + \beta = 28/3$ and $\gamma + \delta = 4$.

Now $(\alpha - \beta)^2 = (\alpha + \beta)^2 - 4\alpha\beta = (28/3)^2 - 12 = 676/9.$

$\therefore$ $\alpha - \beta = \dfrac{26}{3}$ and $\alpha + \beta = \dfrac{28}{3}$

$\Rightarrow$ $\alpha = 9, \alpha = \dfrac{1}{3}.$

Now $(\gamma - \delta)^2 = (\gamma + \delta)^2 - 4\gamma\delta = (4)^2 - 4 \times 3 = 4.$

$\therefore$ $\gamma - \delta = 2$ and $\gamma + \delta = 4$

$\Rightarrow$ $\gamma = 3, \delta = 1.$

Hence the roots are $9, \frac{1}{3}, 3$ and 1.

Example 11:

Solve $6x^3 - 11x^2 - 3x + 2 = 0$, given that roots are in harmonical progression.

Solution:

Let α, β, γ be the roots of the given equation. Then $\alpha + \beta + \gamma = 11/6$, $\alpha\beta + \beta\gamma + \gamma\alpha = -1/2$, $\alpha\beta\gamma = -1/3$.

Since α, β, γ are in H.P., therefore

$$\beta = \frac{2\alpha\gamma}{\alpha + \gamma}.$$

$\Rightarrow$ $\alpha\beta + \beta\gamma + \alpha\gamma = 3\alpha\gamma$

$\Rightarrow$ $-\dfrac{1}{2} = 3\alpha\gamma \Rightarrow \alpha\gamma = -\dfrac{1}{6}.$

Now $\alpha\beta\gamma = -1/3$ and $\alpha\gamma = -1/6 \Rightarrow \beta = 2.$

Thus, 2 is a root of the given equation. Dividing the given equation by $(x - 2)$, we get

$$6x^3 - 11x^2 - 3x + 2 = (x - 2)(6x^2 + x - 1)$$
$$= (x - 2)(3x - 1)(2x + 1).$$

Hence the required roots are $2, \frac{1}{3}, -\frac{1}{2}$

Example 12:

If α, β, γ are the roots of $x^3 + qx + r = 0$, find the equation whose roots are:

1. $\dfrac{\beta^2+\gamma^2}{\alpha}, \dfrac{\gamma^2+\alpha^2}{\beta}, \dfrac{\alpha^2+\beta^2}{\gamma}$.

2. $l\alpha + m\beta\gamma,\ l\beta + m\gamma\alpha,\ l\gamma + m\alpha\beta$.

Solution:

We have $\alpha + \beta + \gamma = 0$, $\alpha\beta + \beta\gamma + \gamma\alpha = q$ and $\alpha\beta r = -r$. ...(1)

Let $y = \dfrac{\beta^2+\gamma^2}{\alpha} \Rightarrow \alpha y + \beta^2 + \gamma^2 \Rightarrow \alpha y = (\alpha + \gamma)^2 - 2\beta\gamma$

$\Rightarrow \quad \alpha y = (-\alpha)^2 - 2(-r/\alpha)$, using (1)

$\Rightarrow \quad \alpha^2 y = \alpha^2 + 2r \Rightarrow \alpha^3 - \alpha^2 y + 2r = 0.$ ...(2)

Since α is a root of $x^3 + qx + r = 0$, therefore

$$\alpha^3 + q\alpha + r = 0. \qquad ...(3)$$

Subtracting (2) from (3), we get

$$y\alpha^2 + q\alpha - r = 0. \qquad ...(4)$$

Multiply (4) by α, (3) by y and subtract.

$$q\alpha^2 - qy\alpha - r(\alpha + y) = 0$$

$\Rightarrow \quad q\alpha^2 - (qy + r)\alpha - ry = 0.$ ...(5)

Solving (4) and (5) for α and α^2, we obtain

$$\frac{\alpha^2}{-qr-(qy+r)r} = \frac{\alpha}{-qr+y^2 r} = \frac{1}{-y(qy+r)-q^2}$$

$$\Rightarrow \quad \alpha^2 = \frac{r[qy+r+q]}{y(qy+r)+q^2},\ \alpha = \frac{r(q-y^2)}{y(qy+r)+q^2}$$

Eliminating α, we get

$$\frac{r(qy+r+q)}{y(qy+r)+q^2} = \frac{r(q-y^2)^2}{[y(qy+r)+q^2]^2}.$$

Hence $[y(qy + r) + q^2](qy + r + q) = r(q - y^2)^2$ is the required equation.

2. Let $y = l\alpha + m\beta\gamma \Rightarrow \alpha y = l\alpha^2 + m\alpha\beta\gamma$

$\Rightarrow \quad \alpha y = l\alpha^2\ mr \Rightarrow l\alpha^2 - y\alpha - mr = 0.$ ...(1)

Since α is a roots of $x^3 + qx + r = 0$, therefore

$$\alpha^3 + q\alpha + r = 0. \quad ...(2)$$

Multiply (1) by α, (2) by l and subtract,

$$y\alpha^2 + m\alpha r + ql\alpha + rl = 0$$

$$\Rightarrow \quad y\alpha^2 + (mr + lq)\,\alpha + lr = 0. \quad ...(3)$$

Solving (1), (3) for α^2 and α, we obtain

$$\frac{\alpha^2}{-lry + nr(mr + lq)} = \frac{\alpha}{-mry + l^2r} = \frac{1}{l(mr + lq) + y^2}$$

$$\therefore \alpha^2 = \frac{mr(mr + lq) - lry}{l(mr + lq) + y^2}, \alpha = \frac{-(mry + l^2r)}{l(mr + lq) + y^2}.$$

Eliminating α, we obtain

$$\frac{r[m(mr + lq) - ly]}{l(mr + lq) + y^2} = \frac{r^2(my + l^2)^2}{[l(mr + lq) + y^2]^2}.$$

Hence $[l(mr + lq) + y^2][m(mr + la) - lv] = (my + l^2)^2$ is the required equation.

Example 13:

If α, β, γ are the roots of $x^3 + qx + r = 0$, form the equation whose roots are

1. $\alpha(\beta + \gamma), \beta(\gamma + \alpha), \gamma(\alpha + \beta)$.
2. $\frac{\beta + \gamma}{\alpha^2}, \frac{\gamma + \alpha}{\beta^2}, \frac{\alpha + \beta}{\gamma^2}$
3. $\frac{1}{\beta} + \frac{1}{\gamma}, \frac{1}{\gamma} + \frac{1}{\alpha}, \frac{1}{\alpha} + \frac{1}{\beta}$.
4. $\beta^2\gamma^2, \gamma^2\alpha^2, \alpha^2\beta^2$.

Solution:

We have $\alpha + \beta + \gamma = 0, \alpha\beta + \beta\gamma + \gamma\alpha = q, \alpha\beta\gamma = -r.$ ...(1)

1. Let $y = \alpha(\beta + \gamma) = \alpha(-\alpha)$ ($\because \alpha + \beta + \gamma = 0$)

$\Rightarrow \quad y = -\alpha^2$. Taking $y = -x^2$, we see that

$$x^2\, qx + qx + r = 0$$

$$\Rightarrow \quad -yx + qx + r = 0$$

$\Rightarrow \quad x(y - q) = r \Rightarrow x^2(y - q)^2 = r^2$

$\Rightarrow \quad -y(y - q)^2 = r^2.$

Hence $y^3 - 2qy^2 + q^2y + r^2 = 0$ is the required equation.

2. Let $y = \dfrac{\beta + \gamma}{\alpha^2}, \dfrac{-\alpha}{\alpha^2} = -\dfrac{1}{\alpha}.$

We take $y = -\dfrac{1}{x} \Rightarrow x = -\dfrac{1}{y}$ in $x^3 + qx + r = 0.$

$\therefore \quad \dfrac{1}{y^2} - \dfrac{q}{y} + r = 0.$ Hence $ry^3 - qy^2 - 1 = 0$ is the required equation.

3. Let $y = \dfrac{1}{\beta} + \dfrac{1}{\gamma} = \dfrac{\beta + \gamma}{\beta\gamma} = -\dfrac{\alpha}{\beta\gamma}$

$\Rightarrow \quad y = \dfrac{-\alpha^2}{\alpha\beta\gamma} = \dfrac{-\alpha^2}{-r} = \dfrac{\alpha^2}{r}$, using (1).

We take $y = \dfrac{x^2}{r} \Rightarrow x^2 = ry$ in $x^3 + qx + r = 0.$

$\therefore \quad (ry)x + qx + r = 0 \Rightarrow x(ry + q) = -r$

$\Rightarrow \quad x^2(ry + q)^2 = r^2 \Rightarrow ry(ry + q)^2 = r^2.$

Hence $y\,(ry + q)^2 = r$ is the required equation.

4. Let $y = \beta^2\gamma^2 \Rightarrow \alpha^2 y = \alpha^2\beta^2\gamma^2 = r^2$ $\qquad (\because \alpha\beta\gamma = -r)$

$\Rightarrow \quad y = r^2/\alpha^2.$ We take $y = r^2/x^2 \Rightarrow x^2 = r^2/y.$

$\therefore \quad x^2 + qx + r = 0$

$\Rightarrow \quad \dfrac{r^2}{y}\,x + qx + r = 0$

$\Rightarrow \quad \left(\dfrac{r^2}{y} + q\right) = -r \Rightarrow x^2\left(\dfrac{r^2}{y} + q\right)^2 = r^2$

$\Rightarrow \quad \dfrac{r^2}{y}\left(\dfrac{r^2}{y} + q\right)^2 = r^2$

$\Rightarrow \quad (r^2 + yq)^2 = y^3.$

Hence $y^3 - q^2y^2 - 2qr^2y - r^4 = 0$ is the required equation.

Example 14:

If the roots of $x^3 + 3px^2 + 3qx + r = 0$ are in harmonical progression, prove that $2q^3 = r\,(3pq - r)$.

Solution:

Let α, β, γ be the roots of $x^3 + 3px^2 + 3px + r = 0$.

$\therefore \alpha + \beta + \gamma = 3p, \alpha\beta + \beta\gamma + \gamma\alpha = 3p, \alpha\beta\gamma = -r.$

Since α, β, γ are in H.P., therefore

$$\beta = \frac{2\alpha\gamma}{\alpha + \gamma} \Rightarrow \alpha\beta + \beta\gamma + \gamma\alpha = 3\gamma\alpha \Rightarrow 3p = 3\gamma\alpha$$

$\Rightarrow \quad q = \gamma\alpha \Rightarrow \alpha\beta = \alpha\beta\gamma = -r$

$\Rightarrow \quad \beta = -r/q.$

Since β is a root of $x^3 + 3px^2 + 3qx + r = 0$, therefore

$$\beta^3 + 3p\beta^2 + 3q\beta + r = 0$$

$\Rightarrow \quad (-r/)^3 + 3p(-r/q)^2 + 3q(-r/q) + r = 0$

$\Rightarrow \quad -r^3 + 3pqr^2 - 3rq^3 + rq^3 = 0.$

Hence $2q^3 = r(3pq - r)$.

Example 15:

If α, β, γ are the roots of $x^3 + qx + r = 0$, form the equation whose root are

$$\beta^2 + \beta\gamma + \gamma^2,\ \gamma^2 + \gamma\alpha + \alpha^2,\ \alpha^2 + \alpha\beta + \beta^2.$$

Solution:

Let $y = \alpha^2 + \alpha\beta + \beta^2 = (\alpha + \beta)^2 - \alpha\beta = (-\gamma)^2 - \alpha\beta$

$\Rightarrow \quad y = \gamma^2 - \alpha\beta \Rightarrow \gamma y = \gamma^3 - \alpha\beta\gamma$

$\Rightarrow \quad \gamma y = \gamma^3 + r$

$\therefore \quad \gamma^2 - \gamma y + r = 0.$ ...(1)

Since γ is a root of $x^3 + qx + r = 0$, therefore

$\gamma^3 + qy + r = 0.$ ...(2)

Subtracting (1) from (2), we get

$y(q + y) = 0 \Rightarrow + y = 0$ $\quad (\because \gamma \neq 0)$

Hence $(q + y)^3 = 0$ is the required equation.

Example 16:

If α, β, γ are the roots of $x^3 + qx + r = 0$, form the equation where roots are

$$\frac{\alpha}{\beta} + \frac{\beta}{\alpha}, \frac{\gamma}{\alpha} + \frac{\alpha}{\gamma}, \frac{\gamma}{\beta} + \frac{\beta}{\gamma}.$$

Solution:

Let $y = \frac{\alpha}{\beta} + \frac{\beta}{\alpha} = \frac{\alpha^2 + \beta^2}{\alpha\beta}$.

We have $\alpha + \beta + \gamma = 0 \Rightarrow \alpha + \beta = -\gamma \Rightarrow (\alpha + \beta)^2 = \gamma^2$

$\Rightarrow \quad \alpha^2 + \beta^2 = \gamma^2 - 2\alpha\beta$. Thus

$$y = \frac{\gamma^2 - 2\alpha\beta}{\alpha\beta} = \frac{\gamma^2}{\alpha\beta} - 2 = \frac{\gamma^3}{\alpha\beta\gamma} - 2 = -\frac{\gamma^3}{r} - 2$$

$\therefore \quad r(y + 2) = -\gamma^2$. We take $x^3 = -r(y + 2)$.

Now $x^3 + qx + r = 0 \Rightarrow -r(y + 2) + qx + r = 0$

$\Rightarrow \quad qx = r(y + 1) \Rightarrow q^3x^3 = r^3(y + 1)^3$

$\Rightarrow \quad rq^3(y + 2) = r^3(y + 1)^3$

$\Rightarrow \quad r^2(y^3 + 1 + 3y^2 + 3y) + q^3(y + 2) = 0.$

Hence $r_2y^3 + 3r^2y^2 + (3r^2 + q^3)y + r^2 + 2p^3 = 0$ is the required equation.

Example 17:

Solve the equation $3x^3 + 11x^2 + 12x + 4 = 0$, being given that the roots are in H.P. **(D.U., B.A. (P) 2000)**

Solution:

The equation whose roots are the reciprocals of the roots of the given equation is given by replacing x by 1/y. The transformed equation is

$$\frac{3}{y^3} + \frac{11}{y^2} + \frac{12}{y} + 4 = 0 \Rightarrow 4y^3 + 12y^2 + 11y + 3 = 0. \qquad ...(1)$$

The three roots (in H.P.) of the given equation are the three roots (in A.P.) of equation (1), say a – d, a, a + d. Then

$$a - d + a + a + d = -3 \Rightarrow a = -1$$

and $\quad a(a - d) + a(a + d) + (a - d)(a + d) = 11/4$

$\Rightarrow 3a^2 - d^2 = 11/4 \Rightarrow d^2 = 1/4 \Rightarrow d = \pm 1/2.$

Taking $a = -1$ and $d = 1/2$, the roots of (1) aer –3/2, –1, –1/2. Hence the roots of the given equation are –2/3, –1, –2.

Example 18:

(1) Solve the equation $x^3 - 6x^2 + 11x - 6 = 0$, the roots being in A.P.

(D.U., Maths (H) 2000)

(2) Solve the equation $x^4 - 12x^3 + 49x^2 - 78 + 40 = 0$, by removing its second term. **(D.U., Maths (H) 2000)**

Solution:

1. Let the three roots be a – d, a, a + d. Then

$$a - d + a + a + d = 6 \Rightarrow a = 2.$$

Also $a(a - d) + a(a + d) + (a - d)(a + d) = 11$

$\Rightarrow \quad 3a^2 - d^2 = 11 \Rightarrow d^2 = 1 \Rightarrow d = \pm 1.$

Hence the required roots are 2 – 2, 2 + 1 *i.e.*, 1, 2, 3.

Example 19:

Solve the equation $27x^3 + 42x^2 - 28x - 8 = 0$, the roots being in G.P.

(D.U., B.A. (P) 1997)

Solution:

Let α/β, α, $\alpha\beta$ be the roots of the given equation. Then

$$\frac{\alpha}{\beta} + \alpha + \alpha\beta = -\frac{42}{27}, \frac{\alpha}{\beta} \cdot \alpha \cdot \alpha B = \frac{8}{27} \Rightarrow \alpha^3 = \frac{8}{27} \Rightarrow \alpha + \frac{2}{3}.$$

$$\therefore \frac{2}{3}\left(\frac{1}{\beta} + 1 + b\right) = -\frac{42}{27} \Rightarrow \beta^2 + \beta + 1 = -\frac{7}{3}\beta \Rightarrow 3\beta^2 + 10\beta + 3 = 0$$

$\Rightarrow \quad (3\beta + 1)(\beta + 3) = 0 \Rightarrow \beta = -1/3, -3.$

Hence the required roots are –2/9, 2/3, –2.

Example 20:

If α, β, γ are the roots of $x^3 + px^2 + qx + r = 0$, form the equation whose roots are $1/\alpha^3$, $1/\beta^3$, $1/\gamma^3$. **(D.U., Maths (H) 1997)**

Solution:

We take $y = 1/x^3$. Dividing the given equation by x^3, we obtain

$$1 + \frac{p}{x} + \frac{q}{x^2} + \frac{r}{x^3} = 0 \Rightarrow 1 + ry = -\frac{1}{x}\left(p + \frac{q}{x}\right) \qquad ...(1)$$

$$\Rightarrow \quad (1 + ry)^3 = -y\left(p + \frac{q}{x}\right)^3 = -y\left[p^3 + \frac{q^3}{x^3} + \frac{2pq}{x}\left(p + \frac{q}{x}\right)\right]$$

$$= y\,[p^3 + q^3y - 3pq\,(1 + ry)], \text{ using (1)}$$

$$\Rightarrow \quad (1 + ry)^3 - 3pqy\,(1 + ry) + y\,(p^3 + q^3y) = 0.$$

Hence $r^3y^3 + (3r^2 - 3pqr + q^3)\,y^2 + (3r - 3pq + p^3)\,y + 1 = 0$

is the required equation.

Example 21:

Solve $27x^4 - 195x^3 - 494x^2 - 520x + 192 = 0$, given that the roots are in geometrical progression.

Solution:

Since the four roots of the given equation are in G.P., we may take them as γ, α, β δ satisfying $\alpha\beta + \gamma\delta$. ...(1)

We have $(\alpha + \beta) + (\gamma + \delta) = \dfrac{195}{27} = \dfrac{65}{9}$ ...(2)

$$(\alpha + \beta)(\gamma + \delta) + \alpha\beta + \gamma\delta = \frac{494}{27} \qquad ...(3)$$

$$(\alpha\beta)(\gamma\delta) = \frac{192}{27} = \frac{64}{9}. \qquad ...(4)$$

From (1) and (4) $\quad \alpha\beta = \dfrac{8}{3} = \gamma\delta.$ ...(5)

From (3) and (5), $(\alpha + \beta)(\gamma + \delta) = \dfrac{494}{27} - \dfrac{16}{3} = \dfrac{350}{27}.$ ...(6)

From (2) and (6), $(\alpha + \beta)$ and $(\gamma + \delta)$ are the roots of

$$y^2 - \frac{65}{9}y + \frac{350}{27} = 0$$

$\Rightarrow 27y^2 - 195y + 350 = 0$

$\Rightarrow \quad (9y - 35)(3y - 10) = 0.$

We take $\alpha + \beta = \dfrac{10}{3}$ and $\gamma + \delta = \dfrac{35}{9}$. ...(7)

Now $(\alpha + \beta)^2 = (\alpha + \beta)^2 - 4\alpha\beta = \dfrac{100}{9} - \dfrac{32}{3} = \dfrac{4}{9} \Rightarrow \alpha - \beta = \dfrac{2}{3}.$ ...(8)

$(\gamma + \delta)^2 = (\gamma + \delta)^2 - 4\gamma\delta = \dfrac{1225}{81} - \dfrac{32}{3} = \dfrac{361}{81} \Rightarrow \gamma - \delta = \dfrac{19}{9}.$ ...(9)

From (7) and (8), $2\beta = 4$

$\Rightarrow \quad \beta = 2$ and $\alpha = \dfrac{10}{3} - 2 = \dfrac{4}{3}.$

From (7) and (9), $2\gamma = 6$

$\Rightarrow \quad \gamma = 3$ and $\delta = \dfrac{35}{9} - 3 = \dfrac{8}{9}.$

Hence the roots are $\frac{8}{9}, \frac{4}{3}$, 2, 3.

Example 22:

Solve $4x^4 - 24x^3 + 31x^2 + 6x - 8 = 0$, being given that the sum of two of its roots is zero.

Solution:

Let the root be α, β, γ, δ, where $\alpha + \beta = 0$. ...(1)

We have $\alpha + \beta + \gamma + \delta = 6$ ...(2)

$$(\alpha + \beta)(\gamma + \delta) + \alpha\beta + \gamma\delta = \frac{31}{4} \quad ...(3)$$

$$\alpha\beta(\gamma + \delta) + \gamma\delta(\alpha + \beta) = -\frac{3}{2} \quad ...(4)$$

$$\alpha\beta\gamma\delta = -2. \quad ...(5)$$

From (1) and (2), $\gamma + \delta = 6$. ...(6)

From (1), (4) and (6); $\alpha\beta = -1/4$. ...(7)

From (1) and (7), $\alpha^2 = 1/4 \Rightarrow \alpha = \pm 1/2$.

From (5) and (7), $\gamma\delta = 8 \Rightarrow \gamma(6 - \gamma) = 8$, by (6)

$\Rightarrow \quad \gamma^3 - 6\gamma + 8 = 0$

$\Rightarrow \quad (\gamma - 4)(\gamma - 2) = 0$

$\Rightarrow \quad \gamma = 2, 4.$

Hence the required roots are $\pm \frac{1}{2}$, 2, 4.

EXERCISES

1. Transform the equation $x^3 - 4x^2 + \frac{1}{4}x - \frac{1}{9} = 0$ into another with integral coefficients and the leading coefficient unity.

   (**Ans.** $x^3 - 24x^2 + 9x - 24 = 0$)

2. Transform the equation $72x^3 - 54x^2 + 45x - 7 = 0$ into another with integral coefficients and unity for the coefficient of the first term.

   (**Ans.** $x^3 - 9x^2 + 90x - 168 = 0$)

   [**Hint:** The given equation is $x^3 - \frac{3}{4}x^3 + \frac{5}{8}x - \frac{7}{72} = 0$.

   Put $y = kx$ *i.e.*, $x = y/k$ and take $k = 2^2 . 3 = 12$]

3. If the sum of two roots of the equation

   $4x^4 - 24x^3 + 31x^2 + 6x - 8 = 0$

   be zero, find all roots of the equation. (**Ans.** $\pm \frac{1}{2}$, 2, 4,)

4. Form an equation whose roots are the reciprocals of the roots of the equation $x^4 - 2x^3 + 3x^2 - 5x + 1 = 0$.

   (**Ans.** $x^4 - 5x^3 + 3x^2 - 2x + 1 = 0$)

   [**Hint:** Take $y = 1/x \Rightarrow x = 1/y$.]

5. Form an equation whose roots are the negatives of the roots of the equation $x^3 - 3x^2 + 5x - 4 = 0$. (**Ans.** $x^3 + 3x^2 + 5x + 4 = 0$)

   [**Hint:** Take $y = -x \Rightarrow x = -y$.]

6. From an equation whose roots are three times those of the equation $x^3 + 2x^2 - 3x + 1 = 0$. (**Ans.** $y^3 + 6y^2 - 27y + 27 = 0$)

   [**Hint:** Take $y = 3x \Rightarrow x = y/3$]

7. If α, β, γ be the roots of $x^3 + px^2 + qx + r = 0$, form the equation whose roots are $\frac{\alpha^2 - \beta\gamma}{\alpha}, \frac{\beta^2 - \gamma\alpha}{\beta}, \frac{\gamma^2 - \alpha\beta}{\gamma}$.

   (**Ans.** $r(y + p)^3 + q^2(y + p)^2 + pq^2(y + p) - q^3 = 0$]

   [**Hint:** Let $y = \frac{1}{\alpha}(\alpha^2 - \beta\gamma) = \alpha = \frac{\alpha\beta\gamma}{\alpha^2} = \alpha + \frac{r}{\alpha^2}$

   $\Rightarrow \quad \alpha^3 - y\alpha^3 + r = 0$. Also $\alpha^3 + p\alpha^2 + q\alpha + r = 0$. ...(1)

   $\therefore \quad \alpha^2(y + p) + q\alpha = 0 \Rightarrow \alpha = -q/(p + y)$. Put it in (1).]

8. If α, β, γ be the roots of $x^3 + px^2 + qx + r = 0$, form the equation whose roots are $\alpha/(\beta + \gamma)$, $\beta/(\gamma + \alpha)$, $\gamma/(\alpha + \beta)$.

   (**Ans.** $(pq - r)y^3 + (2pq - p^3 - 3r)y^3(pq - 3r)y - r = 0$)

   $\left[\right.$**Hint** $y = \frac{\alpha}{\beta + \gamma} = \frac{a}{-\alpha - p}(\because \alpha + \beta + \gamma = -0)$.

   Take $y = \frac{x}{-x - p}$ i.e., $x = -\frac{py}{1 + y}$. $\left.\right]$

9. Find a necessary condition for the roots of the equation $ax^3 + bx^2 + cx + d = 0$ to be in A.P. (**Ans.** $2b^3 - 9abc + 27a^2d = 0$)

10. Solve $4x^3 + 16x^2 - 9x - 36 = 0$, the sum of two of the roots being zero. (**Ans.** $\pm 3/2, -4$)

11. Solve the equation $x^3 - 6x^2 + 11x - 6 = 0$, whose roots are in A.P.
(**Ans.** 1, 2, 3)

12. The root of $3x^3 - x^2 - 3x + 1 = 0$ are in H.P. Find them.
(**Ans.** $-1, 1, \frac{1}{3}$)

13. If α, β, γ be the roots of $x^3 - x + 1 = 0$, find the value of $\alpha^5 + \beta^5 + \gamma^5$.
(**Ans.** -5)

14. If α, β, γ, be the roots of $x^3 + 5x^2 - 6x + 3 = 0$, find the values of (1) $\Sigma\alpha^{-3}$, (2) $\Sigma\alpha^{-4}$ (3) $\Sigma\alpha^3$.
(**Ans.** (1) -3, (2) $-70/9$, (3) -224)

15. If α, β, γ be the roots of $x^3 + qx + r = 0$, find the value of $\Sigma(\alpha - \beta)^2$. (**Ans.** $-6q$)
[**Hint:** $\Sigma(\alpha - \beta)^2 = 2 [(\alpha + \beta + \gamma)^2 - 3 (\alpha\beta + \beta\gamma + \gamma\alpha)]$
$= 2 (0 - 3q = - 6q]$

16. If α, β, γ be the roots of $x^3 + px^2 + qx + r = 0$, find the values of (a) $\Sigma\alpha^{-2}$, (b) $\Sigma\alpha^3$.
(**Ans.** (a) $(q^2 - 2pr)/r^2$, (b) $- p^2 + 3pq - 3r$)
[**Hint:** (a) $\Sigma\alpha^{-2} = 1/\alpha^2 + 1/\beta^2 + 1/\gamma^2 = \dfrac{\sum\alpha^2\beta^2}{\alpha^2\beta^2\gamma^2}$.

17. If $\alpha, \beta, \gamma, \delta$ be the roots of $ax^4 + bx^3 + cx^2 + dx + e = 0$, find the values of (1) $\Sigma\alpha^2\beta$, (2) $\Sigma\alpha^4$.
(**Ans.** (1) $\dfrac{1}{a^3}(b^2c - 2ac^2 - abd + 4ea^2)$
(2) $\dfrac{1}{a^4}(b^4 - 4ab^2c + 2a^2c^2 + 4a^2bd - 4a^2c)$

18. If α, β, γ be the roots of $x^3 + px^2 + qx - r = 0$, find the values of $\Sigma\alpha^2\beta$. (**Ans.** $pq - 3r$)

19. If α, β, γ be the roots of $x^3 + px + qx - r = 0$, form the equation whose roots are: (1) $\beta + \gamma, \gamma + \alpha, \alpha + \beta$, (2) $\beta\gamma/\alpha, \gamma\alpha/\beta\ \alpha\beta/\gamma$.
(**Ans.** (1) $y^3 + qy - r = 0$, (2) $ry^3 + q^2y^2 -2ry + r^2 + 0$)
[**Hint:** (1) $y = \beta + \gamma = - \alpha$ ($\because \alpha + \beta + \gamma = 0$). Take $y = - x$.
(2) $y = \beta\gamma/\alpha^2 = - r/\alpha^2$.
Take $y = - r/x^2 \Rightarrow x^2 = - r/y$.]

20. Form an equation whose roots are the squares of the roots of the equation $3x^2 - 2x^2 + x - 4 = 0$. **(Ans.** $9y^3 + 3y^2 - 15y - 16 = 0$)

21. Find the equation whose roots are the roots of

$$x^4 - 5x^3 + 7x^2 - 17x + 11 = 0,$$

each diminished by 4. **(Ans.** $y^4 + 11y^3 + 43y^2 + 55y - 9 = 0$)

22. Solve the equation $3x^3 + 11x^2 + 12x + 4 = 0$, given that the roots are in H.P.

(Ans. $-1, -2, -2/3$)

[Hint: We have $\alpha\beta + \beta\gamma + \gamma\alpha = 4$ and $\alpha\beta\gamma = -4/3$. Since $\alpha\beta + \beta\gamma + \gamma\alpha = 4/3$ and hence $\beta = -1 \Rightarrow x + 1$ is a factor of the given equation. Now given equation is $(x + 1)(3x^3 + 8x + 4) = 0$. Solving $3x^2 + 8x + 4 = 0$, $x = -2, 2/3$].

2

TRANSFORMATION OF EQUATIONS

2.1 INTRODUCTION

It is sometimes useful to transform an equation into another equation whose roots bear some assigned relations to the roots of the given equation. When the transformed equation is solved, the roots of the given equation can be found by means of the assigned relations. In this chapter, we shall discuss some of the important elementary transformations.

2.2 ROOTS WITH SIGNS CHANGED

To transform an equation into another whose roots of the given equation with their signs changed.

Let $\alpha_1, \alpha_2, \ldots\ldots, \alpha_n$ be the roots of the given equation

$$f(x) \equiv a_0x^n + a_1x^{n-1} + a^2x^{n-2} + \ldots\ldots + a_{n-1}x + a_n = 0. \qquad \ldots(1)$$

Then $a_0x^n + a_1x^{n-1} + a_2x^{n-2} + \ldots\ldots + a_{n-1}x + a_n$

$$\equiv a_0(x - \alpha_1)(x - \alpha_2) \ldots\ldots (x - \alpha_n). \qquad \ldots(2)$$

Now, let y be a root of the required transformed equation. Then $y = -x$.

$\therefore \qquad x = -y.$

Replacing x by $-y$ in the identity (2), we have

$$a_0(-y)^n + a_1(-y)^{n-1} + a_2(-y)^{n-2} + \ldots\ldots + a_{n-1}(-y) + a_n$$
$$\equiv a_1(-y - \alpha_1)(-y - \alpha_2) \ldots\ldots (-y - \alpha_n)$$

or $(-1)^n [a_0y^n - a_1y^{n-1} + a_2y^{n-2} - \ldots\ldots + (-1)^{n-1} a_{n-1}y + (-1)^n a_n]$

$$\equiv (-1)^n [(y + \alpha_1)(y + \alpha_2) \ldots\ldots (y + \alpha_n)]$$

$$a_0y^n - a_1y^{n-1} + a_2y^{n-2} - \ldots\ldots + (-1)^{n-1}a_{n-1}y + (-1)^n a_n$$
$$\equiv (y + \alpha_1)(y + \alpha_2) \ldots\ldots (y + \alpha_n). \quad \ldots(3)$$

From (3), it is clear that $-\alpha_1, -\alpha_2, \ldots\ldots -\alpha_n$ are the roots of the equation

$$a_0y^n - a_1y^{n-1} + a_2y^{n-2} - \ldots\ldots + (-1)^{n-1}a_{n-1}y + (-1)^n a_n = 0.$$

Thus, to get the required transformed equation, we have the following working rule.

Working Rule: *Change the sign of every alternate term of the given equation beginning with the second after making the equation complete if it is not already so.*

2.3 ROOTS MULTIPLIED BY A CONSTANT m

To transform an equation into another whose roots are the roots of the given equation multiplied by a constant m.

Let $\alpha_1, \alpha_2, \ldots\ldots, \alpha_n$ be the roots of the given equation

$$f(x) \equiv a_0x^n + a_1n^{-1} + a_2x^{n-2} + \ldots\ldots + a_{n-1}x + a_n = 0. \quad \ldots(1)$$

Then, $a_0x^n + a_1x^{n-1} + a_2x^{n-2} + \ldots\ldots + a_{n-1}x + a_n$

$$\equiv a_0(x - \alpha_1)(x - \alpha_2) \ldots\ldots (x - \alpha_n). \quad \ldots(2)$$

Now, let y be a root of the required transformed equation.

Then $y = mx$. $\quad \therefore\ x = \dfrac{y}{m}$.

Replacing x by $\dfrac{y}{m}$ in the identity (2), we have

$$a_0\left(\frac{y}{m}\right)^n + a_1\left(\frac{y}{m}\right)^{n-1} + a_2\left(\frac{y}{m}\right)^{n-2} + \ldots\ldots + a_{n-1}\left(\frac{y}{m}\right) + a_n$$
$$\equiv a_0\left(\frac{y}{m} - \alpha_1\right)\left(\frac{y}{m} - \alpha_2\right) \ldots\ldots \left(\frac{y}{m} - \alpha_n\right)$$

Multiplying both sides by m^n, we have

$$a_0y^n + ma_1y^{n-1} + m^2a_2y^{n-2} + \ldots\ldots + m^{n-1}a_{n-1}y + m^n a_n$$
$$\equiv a_0(y - m\alpha_1)(y - m\alpha_2) \ldots\ldots (y - m\alpha_n),$$

which shows that $m\alpha_1, m\alpha_2, \ldots\ldots, m\alpha_n$ are the roots of the equation

$$a_0y^n + ma_1y^{n-1} + m^2a_2y^{n-2} + \ldots\ldots + m^{n-1}a_{n-1} + m^n a_n = 0.$$

Thus, to get the required transformation, we have the following working rule.

Working Rule: *Multiply successive coefficients of the given equation, beginning with the second by m, m^2, m^3, ..., m^n, after making the equation complete if it is not already so.*

This transformation helps us the make the coefficient of the leading term unity when it is not so. It is also helpful for removing fractional coefficients from an equation.

2.4 RECIPROCAL ROOTS

To transform an equation into another whose roots are the reciprocals of the roots of the given equation.

Let $\alpha_1, \alpha_2, \ldots\ldots, \alpha_n$ be the roots of the given equation

$$f(x) \equiv a_0x^n + a_1x^{n-1} + a_2x^{n-2} + \ldots\ldots + a_{n-1}x + a_n = 0. \qquad \ldots(1)$$

Then $a_0x^n + a_1x^{n-1} + a_2x^{n-2} + \ldots\ldots + a_{n-1}x + a_n$

$$\equiv \alpha_0(x - \alpha_1)(x - \alpha_2) \ldots\ldots (x - \alpha_n). \qquad \ldots(2)$$

Now let y be a root of the required transformed equation.

Then $y = \dfrac{1}{x}$. $\quad \therefore\ x = \dfrac{1}{y}$.

Replacing x by $\dfrac{1}{y}$ in the identity (2), we have

$$a_0\left(\frac{1}{y}\right)^n + a_1\left(\frac{1}{y}\right)^{n-1} + a_2\left(\frac{1}{y}\right)^{n-2} + \ldots\ldots + a_{n-1}\left(\frac{1}{y}\right) + a_n$$

$$\equiv a_0\left(\frac{1}{y} - \alpha_1\right)\left(\frac{1}{y} - \alpha_2\right)\ldots\ldots\left(\frac{1}{y} - \alpha_n\right).$$

Multiplying both sides by y^n, we have

$$a_0 + a_1y + a_2y^2 + \ldots\ldots + a_{n-1}y^{n-1} + a_ny^n$$

$$\equiv a_0(1 - \alpha_1y)(1 - \alpha_2y) \ldots\ldots (1 - \alpha_ny)$$

or $a_ny^n + a_{n-1}y^{n-1} + a_{n-2}y^{n-2} + \ldots\ldots + a_1y + a_0$

$$\equiv a_0(-1)^n\alpha_1\alpha_2 \ldots\ldots \alpha_n\left(y - \frac{1}{\alpha_1}\right)\left(y - \frac{1}{\alpha_2}\right)\ldots\ldots\left(y - \frac{1}{\alpha_n}\right),$$

which shows that $\dfrac{1}{\alpha_1}, \dfrac{1}{\alpha_2}, \ldots\ldots, \dfrac{1}{\alpha_n}$ are the roots of the equation

$$a_ny^n + a_{n-1}y^{n-1} + a_{n-2}y^{n-2} + \ldots\ldots + a_1y + a_0 = 0.$$

Thus, to get the required transformation, we have the following working rule.

Working Rule: *Reverse the order of the coefficients of the given equation by taking the last coefficient as the first, last but one as the second and so on after making the given equation complete if it is not already so.*

Or

Replace x by 1/x in the given equation.

2.5 RECIPROCAL EQUATION

Definition: *An equation which remains unaltered by changing x into 1/x is called a reciprocal equation.*

Let the given equation be

$$f(x) \equiv a_0x^n + a_1x^{n-1} + a_2x^{n-2} + \ldots\ldots + a_{n-1}x + a_n = 0.$$

Then, $a_nx^n + a_{n-1}x^{n-1} + a_{n-2}x^{n-2} + \ldots\ldots + a_1x + a_0 = 0$

is an equation whose roots are the reciprocals of the roots of the eqn. (1).

If the equations (1) and (2) are the same, comparing the coefficients of like powers of x in (1) abd (2), we get

$$\frac{a_0}{a_n} = \frac{a_1}{a_{n-1}} = \frac{a_2}{a_{n-2}} = \ldots\ldots = \frac{a_{n-1}}{a_1} = \frac{a_n}{a_0}.$$

Hence, $a_n^2 = a_0^2$ or $a_n = \pm a_0$.

Case I: If $a_n = a_0$, we have

$$a_1 = a_{n-1},\ a_2 = a_{n-2},\ \ldots\ldots,$$

i.e., the coefficients of the terms equidistant from the beginning and the end are equal. Such reciprocal equation are called reciprocal equations of the **first class.**

Case II: If $a_n = -a_0$, we have

$$a_1 = -a_{n-1},\ a_2 = -a_{n-2},\ \ldots\ldots,$$

i.e., the coefficients of the terms equidistant from the beginning and the end are equal in magnitude but opposite in sign. Such reciprocal equations are called reciprocal equations of the *second class.*

2.6 STANDARD FORM OF RECIPROCAL EQUATIONS

If α is a root of a reciprocal equation, then $1/\alpha$ is also its root. Hence the roots of a reciprocal equation occur in pairs like α, $1/\alpha$; β, $1/\beta$ etc. Therefore if a reciprocal equation be of an odd degree, then one of the roots must be its own reciprocal *i.e.*, it must be either + 1 or – 1. If such an equation is of the first class, then -1 is its root. If it belongs to the second class, then 1 is its root. In either case the equation can be depressed by one dimension by dividing f(x) by $x + 1$ or $x - 1$ as the case may be. The depressed equation will be of even degree and of first class.

If the reciprocal equation is of even degree and of second class, then $x^2 - 1$ will be a factor of f(x). In this case the equation can be depressed by two dimensions by dividing f(x) by $x^2 - 1$. The depressed equation will be of even degree and of first class.

Thus, we see that all the reciprocal equations can be reduced to a reciprocal equation of even degree and of the first class known as *'standard form'*. It can be easily seen that the substitution $x + 1/x = y$, reduces the degree of a reciprocal equation of standard form to half its former value.

2.7 TO TRANSFORM AN EQUATION INTO ANOTHER WHOSE ROOTS ARE ANY POWERS OF THE ROOTS OF THE GIVEN EQUATION

Let $\alpha_1, \alpha_2, \ldots\ldots, \alpha_n$ be the roots of the given equation

$$f(x) \equiv a_0x^n + a_1x^{n-1} + a_2x^{n-2} + \ldots\ldots + a_{n-1}x + a_n = 0. \qquad \ldots(1)$$

$$\text{Then, } f(x) \equiv a_0(x - \alpha_1)(x - \alpha_2) \ldots\ldots (x - \alpha_n). \qquad \ldots(2)$$

If we are required to form an equation whose roots are $\alpha_1^m, \alpha_2^m, \ldots\ldots, \alpha_n^m$, then first solve the equation $x^m - 1 = 0$.

Let $\lambda_1, \lambda_2, \ldots\ldots, \lambda_m$ be the roots of the equation (3). Replacing x in the identity (2) by $\lambda_1 x, \lambda_2 x, \ldots\ldots, \lambda_m x$ in succession and multiplying both sides of the m identities thus obtained, we shall get an identity of the form

$$\phi(x^m) \equiv b_0\left(x^m - \alpha_1^m\right)\left(x^m - \alpha_2^m\right)\ldots\ldots\left(x^m - \alpha_n^m\right), \qquad \ldots(4)$$

where b_0 is a constant.

Putting $x^m = y$ in (4), we get $f(y) \equiv b_0\left(y - \alpha_1^m\right)\left(y - \alpha_2^m\right)\ldots\ldots\left(y - \alpha_n^m\right)$.

Thus, $\alpha_1^m, \alpha_2^m, \ldots\ldots, \alpha_n^m$ are the roots of the equation $\phi(y) = 0$.

2.8 TO TRANSFORM AN EQUATION INTO ANOTHER WHOSE ROOTS ARE THE ROOTS OF THE GIVEN EQUATION DIMINISHED BY A CONSTANT h

Let the given be

$$f(x) \equiv a_0x^n + a_1x^{n-1} + a_2x^{n-2} + \ldots\ldots + a_{n-1}x + a_n = 0. \qquad \ldots(1)$$

Let y be a root of the transformed equation. Then

$y = x - h$ or $x = y + h$.

Putting $x = y + h$ in (1), the transformed equation is

$$a_0(y + h)^n + a_1(y + h)^{n-1} + ... + a_{n-1}(y + h) + a_n = 0. \quad ...(2)$$

When simplified and arranged in descending powers of y, suppose the equation (2) takes the form

$$A_0y^n + A_1y^{n-1} + A_2y^{n-2} + ... + A_{n-1}y + A_n = 0. \quad ...(3)$$

Now, since $y = x - h$, we get from (3)

$$A_0(x - h)^n + A_1(x - h)^{n-1} + A_2(x - h)^{n-2} + ...$$
$$+ A_{n-1}(x - h) + A_n = 0. \quad ...(4)$$

Obviously, the L.H.S. of (4) must be identical with the L.H.S. of (1). Hence

$$f(x) \equiv A_0(x - h)^n + A_1(x - h)^{n-1} + A_2(x - h)^{n-2} + ... + A_{n-1}(x - h) + A_n$$
$$\equiv (x - h)\,[A_0(x - h)^{n-1} + A_1(x - h)^{n-2} ...... + A_{n-2}(x - h) + A_{n-1}] + A_n.$$

Thus, A_n is the remainder when f(x) is divided by $x - h$ and the quotient is

$$A_0(x - h)^{n-1} + A_1(x - h)^{n-2} + ... + A_{n-2}(x - h) + A_{n-1}.$$

When this quotient is divided by $x - h$, we get A_{n-1} as the remainder. If we continue this process, we shall get all the coefficients A_n, A_{n-1}, ... A_2, A_1 and the first coefficient A_0 is obviously equal to a_0.

To make the calculation work quicker we should make use of the process of synthetic division after making the equation complete if it is not so already.

Note: To increase the roots of a given equation by h we have simply to diminish the roots by –h.

2.9 REMOVAL OR TERMS

One of the most important applications of the transformation discussed in § 2.8 is to remove a certain specified term from an equation.

If we diminish by h the roots of the equation.

$$f(x) \equiv a_0x^n + a_1x^{n-1} + ... + a_n = 0.$$

then $y = x - h$ or $x = y + h$.

Substituting $x = y + h$ the transformed equation is

$$a_0(y + h)^n + a_1(y + h)^{n-1} + a_2(y + h)^{n-2} + ... + a_n = 0$$

or $$a_0y^n + (na_0h + a_1)y^{n-1}$$
$$+ \left[\frac{n(n-1)}{2!}a_0h^2 + (n-1)a_1h + a_2\right]y^{n-1} + ... = 0.$$

If we want that in the transformed equation the second may be absent, then we should choose h such that the coefficient of y^{n-1} in the transformed equation should be zero. This gives us

$$na_0h + a_1 = 0 \text{ or } h = -\frac{a_1}{na_0}.$$

Hence, the second term of the given equation can be removed by diminishing its roots by $-\frac{a_1}{na_0}$.

Similarly, any other term can be removed, but then h will have to be found an equation of a higher degree.

2.10 TO REDUCE THE CUBIC WITH BINOMIAL COEFFICIENTS *i.e.*, $a_0x^3 + 3a_1x^2 + 3a_2x + a_3 = 0$ TO THE FORM IN WHICH THE SECOND TERM MAY BE WANTING AND THE COEFFICIENT OF THE LEADING TERM BE UNITY AND ALL OTHER COEFFICIENTS INTEGRAL

Let $f(x) \equiv a_0x^3 + 3a_1x^2 + 3a_2x + a_3 = 0$ ...(1)

whose roots are, say α, β, γ.

Let us remove the second term of (1) by diminishing its roots by h. Let y be a root of the transformed equation. Then $y = x - h$ and putting $x = y + h$ in (1), the transformed equation is

$$A_0y^3 + 3A_1y^2 + 3A_2y + A_3 = 0. \qquad ...(2)$$

where $A_0 = a_0$, $A_1 = a_0h + a_1$, $A_2 = a_0h^2 + 2a_1h + a_2$,

and $A_3 = a_0h^3 + 3a_1h^2 + 3a_2h + a_3$.

Since we are to remove the second term, therefore we should choose h such that

$$a_0h + a_1 = 0 \text{ or } h = -\frac{a_1}{a_0}.$$

Putting this value of h in the above relations, we get

$$A_2 = a_0\left(-\frac{a_1}{a_0}\right)^2 + 2a_1\left(-\frac{a_1}{a_0}\right) + a_2 = \frac{a_0a_2 - a_1^2}{a_0} = \frac{H}{a_0}$$

where $H = a_0a_2 - a_1^2$,

$$\text{and } A_3 = a_0\left(-\frac{a_1}{a_0}\right)^3 + 3a_1\left(-\frac{a_1}{a_0}\right)^2 + 3a_2\left(-\frac{a_1}{a_0}\right) + a_3$$

$$= \frac{a_0^2 a_3 - 3a_0 a_1 a_2 + 2a_1^3}{a_0^2} = \frac{G}{a_0^2},$$

where $G = a_0^2 a_3 - 3a_0 a_1 a_2 + 2a_1^3$.

Therefore the transformed equation (2) can be written as

$$a_0 y^3 + \frac{3H}{a_0} y + \frac{G}{a_0^2} = 0$$

or $$y^3 + \frac{3H}{a_0^2} y + \frac{G}{a_0^3} = 0. \quad ...(3)$$

To make all the coefficients integers, we multiply the roots of equation (3) by a_0. The transformed equation is then

$$z^3 + 3Hz + G = 0. \quad ...(4)$$

It is to be noted here that $z = a_0 y = a_0(x - h) = a_0\left(x + \frac{a_1}{a_0}\right) = a_0 x + a_1$.

Therefore if α, β, γ are the roots of (1), the roots of the equation (3) are

$$\frac{a_0\alpha + a_1}{a_0}, \frac{a_0\beta + a_1}{a_0}, \frac{a_0\gamma + a_1}{a_0}$$

and those of (4) are $a_0\alpha + a_1$, $a_0\beta + a_1$, $a_0\gamma + a_1$.

Now, $\alpha + \beta + \gamma = -\frac{3a_1}{a_0}$.

$$\therefore \frac{a_1}{a_0} = -\frac{\alpha + \beta + \gamma}{3}.$$

$$\therefore a_0\alpha + a_1 = a_0\left(\alpha + \frac{a_1}{a_0}\right) = a_0\left(\alpha - \frac{\alpha + \beta + \gamma}{3}\right) = \frac{a_0}{3}(2\alpha - \beta - \gamma).$$

Similarly, $a_0\beta + a_1 = \frac{a_0}{3}(2\beta - \gamma - \alpha)$, and $a_0\gamma + a_1 = \frac{a_0}{3}(2\gamma - \alpha - \beta)$.

Hence, the roots of equation (4), may be written as

$$\frac{a_0}{3}(2\alpha - \beta - \gamma), \frac{a_0}{3}(2\beta - \gamma - \alpha), \frac{a_0}{3}(2\gamma - \alpha - \beta).$$

If we put $z = a_0 x + a_1$ in (4) and simplify, it will be equal to the original cubic multiplied by a_0^2, *i.e.*, $a_0^2 f(x) = 0$. Thus,

$$a_0^2[a_0 x^3 + 3a_1 x^2 + 3a_2 x + a_3] \equiv (a_0 x + a_1)^3 + 3H(a_0 x + a_1) + G. \quad ...(5)$$

2.11 TO REDUCE THE BIQUADRATIC WITH BINOMIAL COEFFICIENTS *i.e.*,

$$a_0x^4 + 4a_1x^3 + 6a_2x^2 + 4a_3x + a_4 = 0$$

To the form in which the second term may be wanting and the coefficient of the leading term be unity and all the other coefficients integral

Let $f(x) \equiv a_0x^4 + 4a_1x^3 + 6a_2x^2 + 4a_3x + a_4 = 0$. ...(1)

Let us remove the second term of (1) by diminishing its roots by h. Let y be a root of the transformed equation. Then $y = x - h$ and the transformed equation is

$$A_0y^4 + 4A_1y^3 + 6A_2y^2 + 4A_3y + A_4 = 0$$

where $A_0 = a_0$, $A_1 = a_0h + a_1$, $A_2 = a_0h^2 + 2a_1h + a_2$,

$A_3 = a_0h^3 + 3a_1h^2 + 3a_2h + a_3$,

and $A_4 = a_0h^4 + 4a_1h^3 + 6a_2h^2 + 4a_3h + a_4$.

As in § 2.10, we get $h = -\dfrac{a_1}{a_0}$.

Therefore, $A_2 = \dfrac{H}{a_0}$, $A_3 = \dfrac{G}{a_0^2}$

and $$A_4 = a_0\left(-\frac{a_1}{a_0}\right)^4 + 4a_1\left(-\frac{a_1}{a_0}\right)^3 + 6a_2\left(-\frac{a_1}{a_0}\right)^2 + 4a_3\left(-\frac{a_1}{a_0}\right) + a_4$$

$$= \frac{a_0^3a_4 - 4a_0^2a_1a_3 + 6a_0a_1^2a_2 - 3a_1^4}{a_0^3}$$

$$= \frac{1}{a_0^3}\left[a_0^2\left(a_0a_4 - 4a_1a_3 + 3a_2^2\right) - 3\left(a_0a_2 - a_1^2\right)^2\right]$$

$$= \frac{1}{a_0^3}\left[a_0^2I - 3H^2\right], \text{ where } I = a_0a_4 - 4a_1a_3 + 3a_2^2.$$

Therefore, the transformed equation (2) may be written as

$$a_0y^4 + \frac{6H}{a_0}y^2 + \frac{4G}{a_0^2} + \frac{a_0^2I - 3H^2}{a_0^3} = 0$$

or $$y^4 + \frac{6H}{a_0^2}y^2 + \frac{4G}{a_0^3}y + \frac{a_0^2I - 3H^2}{a_0^4} = 0. \qquad ...(3)$$

To make all the coefficients integers, we multiply the roots of equation (3) by a_0. The transformed equation is then

$$z^4 + 6Hz^2 + 4Gz + \left(a_0^2 I - 3H^2\right) = 0. \qquad ...(4)$$

As in § 2.10, it may be seen here that if $\alpha, \beta, \gamma, \delta$ are the roots of equation (1), then the roots of the equation (4) may be written as

$$\frac{a_0}{4}(3\alpha - \beta - \gamma - \delta),\ \frac{a_0}{4}(3\beta - \alpha - \gamma - \delta),\ \frac{a_0}{4}(3\gamma - \alpha - \beta - \delta),$$

$$\frac{a_0}{4}(3\delta - \alpha - \beta - \gamma).$$

2.12 THE SYMBOL J

There is another function of the coefficients which greatly facilitates the discussion of the biquadratic equation.

This function is written as $a_0a_2a_4 + 2a_1a_2a_3 - a_0a_3^2 - a_1^2a_4 - a_2^3$

and is denoted by the symbol J.

In the form of a determinant, J can be expressed as

$$J = \begin{vmatrix} a_0 & a_1 & a_2 \\ a_1 & a_2 & a_3 \\ a_2 & a_3 & a_4 \end{vmatrix}.$$

The symbols G, H, I, J are connected by a very important relation, *i.e.*,

$$G^2 + 4H^3 = a_0^2(HI - a_0J).$$

This relation can be easily verified.

2.13 TRANSFORMATION IN GENERAL

To form an equation whose roots are the symmetric functions of the roots of a given equation.

Let $\quad f(x) = 0$

be a given equation.

Let y be a root of the transformed equation. Suppose x and y are connected by a relation of the form

$$\phi(x, y) = 0$$

Eliminating x between (1) and (2), we get an equation in y, which will be the required equation. The process will be clear from the following examples.

2.14 EQUATION OF SQUARED DIFFERENCES OF A CUBIC

If α, β, γ be the roots of the cubic $a_0x^3 + 3a_1x^2 + 3a_2x + a_3 = 0$, to form the equation whose roots are $(\alpha - \beta)^2, (\beta - \gamma)^2, (\gamma - \alpha)^2$ and to discuss the nature of the roots of the given cubic.

Let us remove the second term from the given equation

$$a_0x^3 + 3a_1x^2 + 3a_2x + a_3 = 0, \qquad ...(1)$$

by diminishing is roots by h, where $h = -\dfrac{a_1}{a_0}$.

The transformed equation is then

$$y^3 + \frac{3H}{a_0^2}y + \frac{G}{a_0^3} = 0 \quad y^3 + qy + r = 0, \qquad ...(2)$$

where $q = \dfrac{3H}{a_0^2}$ and $r = \dfrac{G}{a_0^3}$.

The roots of the equation (2) are $\alpha - h, \beta - h, \gamma - h$, say α', β', γ'.

$\therefore \quad (\alpha - \beta)^2 = [(\alpha - h) - (\beta - h)]^2 = (\alpha' - \beta')^2.$

Similarly, $(\beta - \gamma)^2 = (\beta' - \gamma')^2$ and $(\gamma - \alpha)^2 = (\gamma' - \alpha')^2$.

Hence, the equation of the squared differences of (1) is the same as that of (2).

Thus, if $\alpha', \beta' \gamma'$ be the roots of the equation (2), we are to find the equation roots are

$$(\alpha' - \beta')^2, (\beta' - \gamma')^2, (\gamma' - \alpha')^2.$$

Let $\quad z = (\alpha' - \beta')^2 = (\alpha' + \beta')^2 - 4\alpha'\beta' = \gamma'^2 - 4\alpha'\beta'$,

since $\quad \alpha' + \beta' + \gamma' = 0.$

$$\therefore \quad z = \gamma'^2 - \frac{4\alpha'\beta'\gamma'}{\gamma'} = \gamma'^2 + \frac{4r}{\gamma'}, \text{ since } \alpha'\beta'\gamma' = -r.$$

$$\therefore \quad \gamma'^3 - \gamma'z + 4r = 0.$$

But γ' is a root of the equation (2). Therefore

$$\gamma'^3 + q\gamma' + r = 0. \qquad ...(3)$$

Subtracting (3) from (4), we get

$$\gamma'(q + z) - 3r = 0 \text{ or } \gamma' = \frac{3r}{q+z}.$$

Putting the value of γ' in (4), we get

$$\left(\frac{3r}{q+z}\right)^3 + q\left(\frac{3r}{q+z}\right) + r = 0$$

or $$(q + z)^3 + 3q(q + z)^2 + 27r^2 = 0 \qquad ...(4)$$

or $$z + 6qz^2 + 9q^2z + (27r^2 + 4q^3) = 0. \qquad ...(5)$$

Putting the values of q and r in (5), we get

$$z^3 + \frac{18H}{a_0^2}z^2 + \frac{81H^2}{a_0^4}z + \frac{27}{a_0^6}\left(G^2 + 4H^3\right) = 0. \qquad ...(6)$$

Thus, $(\alpha - \beta)^2$, $(\beta - \gamma)^2$, $(\gamma - \alpha)^2$ are the roots of the equation (6).

We have from (6),

$$(\alpha - \beta)^2 (\beta - \gamma)^2 (\gamma - \alpha)^2 = -\frac{27}{a_0^6}\left(G^2 + 4H^3\right). \qquad ...(7)$$

Nature of the Roots of the Given Cubic

With the help of the equation of the squared differences of the cubic, we are able to discuss the nature of the roots of the cubic. The coefficients of the cubic are assumed to be real quantities. We know that, complex roots occur in conjugate pairs. Therefore at least one of the roots of the given cubic, say α, must be real. If the other two roots β and γ are also real then the product

$$(\alpha - \beta)^2 (\beta - \gamma)^2 (\gamma - \alpha)^2$$

will be definitely a positive quantity.

If the other two roots β and γ are conjugate complex quantities, let us suppose that $\beta = m + in$ and $\gamma = m - in$, where m and n are real.

Then $(\alpha - \beta)^2 (\beta - \gamma)^2 (\gamma - \alpha)^2 = [\alpha - m - in]^2 [2in]^2 [m - in - \alpha]^2$

$= -4n^2[\alpha - m - in]^2 [\alpha - m + in]^2 = -4n^2[(\alpha - m)^2 + n^2]$

= a negative quantity.

Hence we conclude that if $(\alpha - \beta)^2 (\beta - \gamma)^2 (\gamma - \alpha)^2$ is positive, all the three roots of the cubic must be real and if it is negative the cubic must have two imaginary roots.

Let us now discuss the various cases as below:

1. *When $G^2 + 4H^3$ is negative, the roots of the cubic are all real:* When $G^2 + 4H^3$ is a negative quantity, we have from (7)

$$(\beta - \gamma)^2 (\gamma - \alpha)^2 (\alpha - \beta)^2 = \text{a positive quantity.}$$

Hence all the three roots of the cubic must be real.

2. *When $G^2 + 4H^3 > 0$, the cubic has a pair or imaginary roots:* When $G^2 + 4H^3 > 0$, we have from (7)

$$(\beta - \gamma)^2 (\gamma - \alpha)^2 (\alpha - \beta)^2 = \text{a negative quantity.}$$

Hence the cubic must have a pair of imaginary roots.

3. *When $G^2 + 4H^3 = 0$, the cubic has two equal roots:* When $G^2 + 4H^3 = 0$, we have from (7)

$$(\beta - \gamma)^2 (\gamma - \alpha)^2 (\alpha - \beta)^2 = 0.$$

Now this product of the three factors can be zero only if at least one of the factors, say $(\alpha - \beta)^2$, is equal to zero. Hence in this case two of the roots of the cubic are equal.

4. *When $G = 0$ and $H = 0$, all the roots of the cubic are equal:* In this case, the equation (6) reduces to $z^3 = 0$. It follows that all the roots of (6), viz

$$(\beta - \gamma)^2, (\gamma - \alpha)^2, (\alpha - \beta)^2,$$

are zero, so that $\alpha = \beta = \gamma$.

Note: We have $G^2 + 4H^3 = a_0^2(HI - a_0J) = a_0^2\Delta$,

where $\Delta = HI - a_0J$.

Δ is called the discrimination of the cubic (1). Its vanishing is the necessary and sufficient condition for the cubic to have two equal roots.

SOLVED EXAMPLES

Example 1:

Form an equation whose roots are the reciprocal of the roots of the equation

$$x^4 - 4x^3 + 5x^2 + 11x - 12 = 0.$$

Solution:

The transformed equation is

$$-12x^4 + 11x^3 + 5x^2 - 4x + 1 = 0 \text{ or } 12x^4 - 11x^3 - 5x^2 + 4x - 1 = 0.$$

Example 2:

Transform the equation $72x^3 - 54x^2 + 45x - 7 = 0$ into another with integral coefficients and unity for the coefficient of the first term.

Solution:

The given equation can be written as

$$x^3 - \frac{3}{4}x^2 + \frac{5}{8}x - \frac{7}{72} = 0.$$

The L.C.M. of the numbers 4, 8, 72 in the denominators is 72. Therefore the fractional coefficients can be removed by multiplying the roots of the equation (1) by 72. But a number much smaller than 72 will suffice in this case.

Let the roots of the equation (1) be multiplied by m. Then the transformed equation is

$$x^3 - \frac{3}{4}mx^2 + \frac{5}{8}m^2x - \frac{7}{72}m^3 = 0. \qquad ...(2)$$

Splitting the denominators of the fractional coefficients into smaller factors, the equation (2) can be written as

$$x^3 - \frac{3}{2^2}mx^2 + \frac{5}{2^3}m^2x - \frac{7}{2^3 \cdot 3^2}m^3 = 0.$$

Clearly, the least value of m to remove the fractional coefficients is

$$m = 2^2 \times 3 = 12.$$

So putting $m = 2^2 \cdot 3$ in (3), the required equation is

$$x^3 - \frac{3}{2^2}\cdot(2^2 \cdot 3)x^2 + \frac{5}{2^3}(2^2 \cdot 3)^2 x - \frac{7}{2^3 \cdot 3^2}(2^2 \cdot 3)^3 = 0$$

or $\quad x^3 - 9x^2 + 90x - 168 = 0.$

Example 3:

Change the signs of the roots of the equation

$$2x^5 + 4x^3 - 13x^2 + 7x + 6 = 0.$$

Solution:

Supplying the missing terms with zero coefficients, the corresponding complete equation is

$$2x^5 + 0.x^4 + 4x^3 - 13x^2 + 7x + 6 = 0. \qquad ...(1)$$

Changing the sign of the every alternate term of the equation (1) beginning with the second, the required equation is

$$2x^5 - 0.x^4 + 4x^3 + 13x^2 + 7x - 6 = 0$$

or $\quad 2x^5 + 4x^3 + 13x^2 + 7x - 6 = 0.$

Example 4:

Find the equation whose roots are the reciprocals of the roots of the equation

$$x^4 - 5x^3 + 7x^2 + 3x - 7 = 0.$$

Solution:

The given equation is

$$x^4 - 5x^3 + 7x^2 + 3x - 7 = 0.$$

Replacing x by 1/x, the required equation is

$$\left(\frac{1}{x}\right)^4 - 5\left(\frac{1}{x}\right)^3 + 7\left(\frac{1}{x}\right)^2 + 3\left(\frac{1}{x}\right) - 7 = 0$$

or $$1 - 5x + 7x^2 + 3x^3 - 7x^4 = 0$$

or $$7x^4 - 3x^3 - 7x^2 + 5x - 1 = 0.$$

Example 5:

If α, β, γ, δ be the roots of the equation

$$x^4 + px^3 + qx^2 + rx + s = 0,$$

form an equation whose roots are α^2, β^2, γ^2, δ^2. Hence, find the value of $\Sigma\, \alpha^2\beta^2$.

Solution:

The given equation is

$$x^4 + px^3 + qx^2 + rx + s = 0. \qquad ...(1)$$

Keeping the even degree terms on one side and the rest on the other side, the equation (1) can be written as

$$x^4 + qx^2 + s = -(px^3 + rx)$$

or $$x^4 + qx^2 + s = -x(px^2 + r).$$

Squaring both sides, we get

$$(x^4 + qx^2 + s)^2 = x^2(px^2 + r)^2.$$

Replacing x^2 by y, we get the required equation as

$$(y^2 + qy + s)^2 = y\,(py + r)^2$$

or $y^4 + (2q - p)\,p^3 + (q^2 + 2s - 2pr)\,y^2 + (2qs - r^2)\,y + s^2 = 0$ ...(2)

Now α^2, β^2, γ^2, δ^2 are the roots of the equation (2),

therefore $\Sigma\, \alpha^2\, \beta^2 = q^2 + 2s - 2pr.$

Example 6:

Find the equation whose roots are the cubes of the roots of the equation

$$x^n + p_1x^{n-1} + p_2x^{n-2} + \ldots\ldots + p_{n-1}x + p_n = 0.$$

Solution:

Let $\alpha_1, \alpha_2, \ldots\ldots, \alpha_n$ be the roots of the given equation, then

$(x - \alpha_1)(x - \alpha_2) \ldots\ldots (x - \alpha_n) \equiv x^n + p_1x^{n-1} + p_2x^{n-2} \ldots\ldots + p_{n-2}x^2 + p_{n-1}x + p_n$

$\equiv (p_n + p_{n-3}x^3 + \ldots\ldots) + x(p_{n-1} + p_{n-4}x^3 + \ldots\ldots) + x^2(p_{n-2} + p_{n-5}x^3 + \ldots\ldots)$

$\equiv P + Qx + Rx^2$ ...(1)

where, P, Q, R are polynomials in x which involve powers of x that are integral multiples of 3 only.

Now the roots of the equation $x^3 - 1 = 0$ are $1, \omega, \omega^2$ where $\omega^3 = 1$, $\omega^4 = \omega$, $1 + \omega + \omega^2 = 0$ etc.

Replacing x by $1 \cdot x$,. $\omega \cdot x$ and $\omega^2 x$ in both sides of (1), ω get

$P + Qx + Rx^2 \equiv (x - \alpha_1)(x - \alpha_2) \ldots\ldots (x - \alpha_n)$ ...(2)

$P + Q\omega x + R\omega^2x^2 \equiv (\omega x - \alpha_1)(\omega x - \alpha_2) \ldots\ldots (\omega x - \alpha_n)$ ...(3)

$P + Q\omega^2x + R\omega x^2 \equiv (\omega^2x - \alpha_1)(\omega^2x - \alpha_2) \ldots\ldots (\omega^2x - \alpha_n)$ ...(4)

Multiplying the corresponding sides of (2), (3) and (4), we get

$$P^3 + Q^3x^3 + R^3x^6 - 3PQRx^3 \equiv \left(y - \alpha_1^3\right)\left(y - \alpha_2^3\right) \ldots\ldots \left(y - \alpha_n^3\right).$$

Thus $\alpha_1^3, \alpha_2^3, \ldots\ldots, \alpha_n^2$ are the roots of the equation

$$P^3 + Q^3y + R^3y^2 - 3PQRy = 0.$$

Example 7:

If α, β, γ be the roots of the equation $x^3 + px^2 + qx + r = 0$, find the equation whose roots are $\alpha^3, \beta^3, \gamma^3$. Hence, find the value of $\Sigma\alpha^3$ and $\Sigma\,\alpha^3\beta^3$.

Solution:

The given equation can be written as

$$x^3 + r = -x(px + q). \quad \ldots(1)$$

Let y be a root of the transformed equation, then $y = x^3$.

$\therefore\ x = y^{1/3}$.

Putting $x = y^{1/3}$ in (1), we get

$$y + r = -y^{1/3}(py^{1/3} + q). \quad ...(2)$$

Cubing both sides of (2), we get

$$(y + r)^3 = -y[p^3y + q^3 + 3pqy^{1/3}(py^{1/3} + q)]$$

or $$(y + r)^3 = -y[p^3y + q^3 - 3pq(y + r)],$$

on putting the value of $y^{1/3}(py^{1/3} + q)$ from (2).

Simplifying, we get the required equation as

$$y^3 + (p^3 - 3pq + 3r)y^2 + (q^3 - 3pqr + 3r^2)y + r^3 = 0.$$

Now the roots of the equation (3) are $\alpha^3, \beta^3, \gamma^3$.

$\therefore$ $$\Sigma\alpha^3 = -(p^3 - 3pq + 3r)$$

and $$\Sigma\alpha^3\beta^3 = q^3 - 3pqr + 3r^2.$$

Example 8:

Form the equation whose roots are the squares of the roots of the equation

$$x^n + p_1x^{n-1} + p_2x^{n-2} + \ldots\ldots + p_{n-1}x + p_n = 0.$$

Solution:

Let $\alpha_1, \alpha_2, \ldots\ldots \alpha_n$ be the roots of the given equation, then

$$x^n + p_1x^{n-1} + \ldots\ldots + p_{n-1}x + p_n \equiv (x - \alpha_1)(x - \alpha_2)\ldots\ldots(x - \alpha_n) \quad ...(1)$$

Now the roots of the equation $x^2 - 1 = 0$ are $1, -1$.

Therefore replacing x by $1 . x$ and $-1 . x$ in both sides of (1), we get

$$(x^n + p_2x^{n-2} + \ldots\ldots) + (p_1x^{n-1} + p_3x^{n-3} + \ldots\ldots)$$
$$\equiv (x - \alpha_1)(x - \alpha_2)\ldots\ldots(x - \alpha_n) \quad ...(2)$$

and $$(-1)^n[(x^n + p_2x^{n-2} + \ldots\ldots) - (p_1x^{n-1} + p_3x^{n-3} + \ldots\ldots)]$$
$$\equiv (-1)^n[(x + \alpha_1)(x + \alpha_2)\ldots\ldots(x + \alpha_n)] \quad ...(3)$$

Multiplying both members of the identities (2) and (3), we get

$$(x^n + p_2x^{n-2} + p_4x^{n-4} + \ldots\ldots)^2 - (p_1x^{n-1} + p_3x^{n-3} + \ldots\ldots)^2$$
$$\equiv \left(x^2 - \alpha_1^2\right)\left(x^2 - \alpha_2^2\right)\ldots\ldots\left(x^2 - \alpha_n^2\right)$$

or $$x^{2n} + \left(2p_2 - p_1^2\right)x^{2n-2} + \left(p_2^2 - 2p_1p_3 + 2p_4\right)x^{2n-4} + \ldots\ldots$$
$$\equiv \left(x^2 - \alpha_1^2\right)\left(x^2 - \alpha_2^2\right)\ldots\ldots\left(x^2 - \alpha_n^2\right) \quad ...(4)$$

Replacing x^2 by y in the identity (4), we get

$$y^n+(2p_2-p_1^2)y^{n-1}+(p_2^2-2p_1p_3+2p_4)y^{n-2}+\ldots\ldots$$

$$\equiv(y-\alpha_1^2)(y-\alpha_2^2)\ldots\ldots(y-\alpha_n^2).$$

Thus, $\alpha_1^2, \alpha_2^2, \ldots\ldots, \alpha_n^2$ are the roots of the equation

$$y^n+(2p_2-p_1^2)y^{n-1}+(p_2^2-2p_1p_3+2p_4)y^{n-2}+\ldots\ldots=0.$$

Example 9:

Reduce the equation

$4x^4 - 85x^3 + 357x^2 - 340x + 64 = 0$

to a reciprocal equation, and solve it.

Solution:

The given equation is

$$4x^4 - 85x^3 + 357x^2 - 340x + 64 = 0. \qquad \ldots(1)$$

To reduce the equation (1) to a reciprocal equation, we first multiply the roots of (1) by m and get the following transformed equation

$$4y^4 - 85my^3 + 357m^2y^2 - 340m^3y + 64m^4 = 0,$$

where $y = mx$.

If (2) is to be a reciprocal equation, we have

$$\frac{4}{64m^4}=\frac{85m}{340m^3}$$

or $m^2=\frac{1}{4}$. $\quad\therefore\ m=\pm\frac{1}{2}$.

Taking $m=\frac{1}{2}$ and substituting in (2), the required reciprocal equation is

$$4y^4-\frac{85}{2}y^3+\frac{357}{4}y^2-\frac{340}{8}y+\frac{64}{16}=0$$

or $16y^4 - 170y^3 + 357y^2 - 170y + 16 = 0$

or $16(y^4 + 1) - 170(y^3 + y) + 357y^2 = 0$.

Dividing throughout by y^2, we have

$$16\left(y^2+\frac{1}{y^2}\right)-170\left(y+\frac{1}{y}\right)+357=0$$

or $$16\left\{\left(y+\frac{1}{y}\right)^2-2\right\}-170\left(y+\frac{1}{y}\right)+357=0.$$

Putting $y+\frac{1}{y}=z$, we get

$$16z^2 - 170z + 325 = 0.$$

$$\therefore \quad z=\frac{170 \pm \sqrt{\{(170)^2-4\times16\times325\}}}{2\times16}=\frac{170\pm90}{32}$$

or $$z=\frac{65}{8}, \frac{5}{2}.$$

When $z=\frac{65}{8}$, we have $y+\frac{1}{y}=\frac{65}{8}$

or $$8y^2 - 65y + 8 = 0 \text{ or } (y - 8)(8y - 1) = 0.$$

$$\therefore \quad y = 8, \frac{1}{8}.$$

Again when $z=\frac{5}{2}$, we have

$$y+\frac{1}{y}=\frac{5}{2} \text{ or } 2y^2 - 5y + 2 = 0$$

or $$(y - 2)(2y - 1) = 0.$$

$$\therefore \quad y = 2, \frac{1}{2}.$$

$\therefore$ The roots of the equation (3) are $y = 2, 8, \frac{1}{2}, \frac{1}{8}$.

$$\because \quad y = mx = \frac{1}{2}x,$$

$$\therefore \quad x = 2y.$$

Hence, the roots of the given equation (1) are 4, 16, 1, $\frac{1}{4}$.

Example 10:

Solve the equation $x^5 - 5x^4 + 9x^3 - 9x^2 + 5x - 1 = 0$.

Solution:

The given equation is

$x^5 - 5x^4 + 9x^3 - 9x^2 + 5x - 1 = 0$

or $(x^5 - 1) - 5x(x^3 - 1) + 9x^2(x - 1) = 0$

or $(x - 1)(x^4 + x^3 + x^2 + x + 1) - 5x(x - 1)(x^2 + x + 1) + 9x^2(x-1) = 0$

or $(x - 1)[x^4 + x^3 + x^2 + x + 1 - 5x(x^2 + x + 1) + 9x^2] = 0$

or $(x - 1)(x^4 - 4x^3 + 5x^2 - 4x + 1) = 0.$

$\therefore$ $x - 1 = 0$, giving $x = 1$

or $x^4 - 4x^3 + 5x^2 - 4x + 1 = 0.$ ...(1)

Equation (1) is a reciprocal equation, and can be written as

$(x^4 + 1) - 4(x^3 + x) + 5x^2 = 0.$

Dividing by x^2, we have

$$\left(x^2 + \frac{1}{x^2}\right) - 4\left(x + \frac{1}{x}\right) + 5 = 0$$

or $$\left\{\left(x + \frac{1}{x}\right)^2 - 2\right\} - 4\left(x + \frac{1}{x}\right) + 5 = 0.$$

Putting $x + \frac{1}{x} = y$, we have $y^2 - 4y + 3 = 0$

or $(y - 1)(y - 3) = 0.$

$\therefore$ $y = 1$ or 3.

When $y = 1$, we have $x + \frac{1}{x} = 1$

or $x^2 - x + 1 = 0.$

$$\therefore\ x = \frac{1 \pm \sqrt{(1-4)}}{2} = \frac{1}{2}\left(1 \pm i\sqrt{3}\right).$$

Again when $y = 3$, we have $x + \frac{1}{x} = 3$

or $x^2 - 3x + 1 = 0.$

$$\therefore\ x = \frac{3 \pm \sqrt{(9-4)}}{2} = \frac{1}{2}\left(3 \pm \sqrt{5}\right).$$

Hence, the roots of the given equation are

$$1, \frac{1}{2}\left(1 \pm i\sqrt{3}\right), \frac{1}{2}\left(3 \pm \sqrt{5}\right).$$

Example 11:

Find the equation whose roots are the roots of $x^4 - 5x^3 + 7x^2 - 17x + 11 = 0$ *each diminished by 4.*

Solution:

Let the transformed equation be

$$A_0y^4 + A_1y^3 + A_2y^2 + A_3y + A_4 = 0, \qquad ...(1)$$

where $A_0 = 1$ and A_4, A_3, A_2, A_1 are calculated below by repeatedly dividing

$$x^4 - 5x^3 + 7x^2 - 17x + 11$$

by $x - 4$ by the process of synthetic division.

4	1	– 5	7	– 17	11
		4	– 4	12	– 20
		– 1	3	– 5	– 9, = A_4
		4	12	60	
		3	15	55 = A_3	
		4	28		
		7	43 = A_2		
		4			
		11 = A_1			

Hence, the transformed equation is

$$y^4 + 11y^3 + 43y^2 + 55y - 9 = 0.$$

Explanation: The coefficients of the given polynomial are written in the first line. Division by $x - 4$ gives $- 9$ as the remainder and $x^3 - x^2 + 3x - 5$ as the quotient. Division by $x - 4$ again gives 55 as the remainder and $x^2 + 3x + 15$ as the quotient, and so on. The successive remainders $- 9$, 55, 43, 11 are the values of A_4, A_3, A_2, A_1 respectively.

Example 12:

Transform the equation $x^4 + 8x^3 + x - 5 = 0$ into one in which the second term is wanting.

Solution:

Supplying the missing terms with zero coefficients, the corresponding complete equation is

$$x^4 + 8x^3 + 0x^2 + x - 5 = 0. \qquad ...(1)$$

In order to remove the second term in the equation (1) the roots of the equation are to be diminished by h, given by

$$h = -\frac{a_1}{na_0} = -\frac{8}{4\times 1} = -2.$$

Diminishing the roots of (1) by – 2, let the transformed equation be

$$A_0x^4 + A_1x^3 + A_2x^2 + A_3x + A_4 = 0, \qquad ...(2)$$

where $A_0 = 1$, and A_4, A_3, A_2, A_1 are calculated by dividing the left hand side of the equation (1) by x – (– 2) by the process of synthetic division as shown below:

2	1	8	0	1	– 5
		– 2	– 12	24	– 50
		6	– 12	25	– 55 = A_4
		– 2	– 8	40	
		4	– 20	65 = A_3	
		– 2	– 4		
		2	– 24 = A_2		
		– 2			
	0 = A_1				

Hence from (2), the required transformed equation is

$$x^4 - 24x^2 + 65x - 55 = 0.$$

Example 13:

If α, β, γ be the roots of the equation

$$x^3 + px^2 + qx + r = 0,$$

find the equation whose roots are

$$\frac{\alpha}{\beta+\gamma-\alpha}, \frac{\beta}{\gamma+\alpha-\beta}, \frac{\gamma}{\alpha+\beta-\gamma}.$$

Solution:

Let y be a root of the transformed equation. Then

$$y = \frac{\alpha}{\beta+\gamma-\alpha} = \frac{\alpha}{(\alpha+\beta+\gamma)-2\alpha} \text{ or}$$

$$y = \frac{\alpha}{-p-2\alpha}. \qquad [\because \alpha + \beta + \gamma = -p]$$

$$\therefore -y(p + 2\alpha) = a \text{ or } \alpha = -\frac{py}{1+2y}.$$

But α is a root of the given equation. Hence this value of α should satisfy the given equation. Therefore the transformed equation is

$$\left(-\frac{py}{1+2y}\right)^3 + p\left(-\frac{py}{1+2y}\right)^2 + q\left(-\frac{py}{1+2y}\right) + r = 0$$

or $-p^3y^3 + p^3y^2(1+2y) - pqy(1+2y)^2 + r(1+2y)^3 = 0$

or $(p^3 - 4pq + 8r)y^3 + (p^3 - 4pq + 12r)y^2 + (6r - pq)y + r = 0$.

Example 14:

If α, β, γ are the roots of the equation $x^3 - x - 1 = 0$, find the equation whose roots are

$$\frac{1+\alpha}{1-\alpha}, \frac{1+\beta}{1-\beta}, \frac{1+\gamma}{1-\gamma}.$$

Hence, write down the value of $\Sigma\left(\frac{1+\alpha}{1-\alpha}\right)$.

Solution:

The given equation is $x^3 - x - 1 = 0$.

If y is a root of the required equation, then $y = \frac{1+\alpha}{1-\alpha}$. ...(1)

Replacing α by x since x is a root of the given equation (1), we have

$$y = \frac{1+x}{1-x}. \quad \therefore \quad x = \frac{y-1}{y+1}. \qquad ...(2)$$

Eliminating x between (1) and (2), we get

$$\left(\frac{y-1}{y+1}\right)^3 - \left(\frac{y-1}{y+1}\right) - 1 = 0$$

or $\quad (y-1)^3 - (y-1)(y+1)^2 - (y+1)^3 = 0$

or $\quad y^3 + 7y^2 - y + 1 = 0$, ...(3)

which is the required equation whose roots are

$$\frac{1+\alpha}{1-\alpha}, \frac{1+\beta}{1-\beta}, \frac{1+\gamma}{1-\gamma}.$$

Now the sum of the roots of (3)

$$= \Sigma\left(\frac{1+\alpha}{1-\alpha}\right) = -\frac{\text{coefficient of } y^2}{\text{coefficient of } y^3} = -\frac{7}{1} = -7.$$

Example 15:

If α, β, γ are the roots of the cubic $x^3 - px^2 + qx - r = 0$, form the equation whose roots are

$$\beta\gamma + \frac{1}{\alpha}, \gamma\alpha + \frac{1}{\beta}, \alpha\beta + \frac{1}{\gamma}.$$ **(Meerut, 1992)**

Solution:

The given equation is $x^3 - px^2 + qx - r = 0$,

whose roots are α, β, γ. We have $\alpha\beta\gamma = r$. ...(1)

Let y be a root of the required equation. Then

$$y = \beta\gamma + \frac{1}{\alpha} = \frac{\alpha\beta\gamma + 1}{\alpha} = \frac{r+1}{x}.$$

$\left[\because \alpha\beta\gamma = (-1)^3 . \frac{-r}{1} = r \text{ and } \alpha \text{ may be replaced by } x \text{ which is a root of } (1).\right]$

$$\therefore \ x = \frac{r+1}{y}. \qquad ...(2)$$

Eliminating x between (1) and (2), we get

$$\left(\frac{r+1}{y}\right)^3 - p\left(\frac{r+1}{y}\right)^2 + q\left(\frac{r+1}{y}\right) - r = 0$$

or $\quad ry^3 - q(r+1)y^2 + p(r+1)^2y - (r+1)^3 = 0,$ ...(3)

which is the required equation.

Example 16:

If α, β, γ are the roots of the equation $x^3 + px^2 + qx + r = 0$, find the equation whose roots are $(\beta + \gamma), (\gamma + \alpha), (\alpha + \beta)$ and hence or otherwise find the value of

(i) $\Sigma (\beta + \gamma)(\gamma + \alpha)$ and

(ii) $(\alpha + \beta)(\beta + \gamma)(\gamma + \alpha)$.

Solution:

The given equation is $x^3 + px^2 + qx + r = 0$, ...(1)

whose roots are α, β, γ.

$\therefore \quad \alpha + \beta + \gamma = -p.$

If y is a root of the required equation, then

$$y = \beta + \gamma = (\alpha + \beta + \gamma) - \alpha = -p - \alpha = -p - x$$

[replacing α by x which is the root of (1)]

or $\quad x = -(y + p).$ ...(2)

Eliminating x between (1) and (2), we get

$$-(y + p)^3 + p(y + p)^2 - q(y + p) + 1 = 0$$

or $\quad y^3 + 2py^2 + (p^2 + q)y + (pq - r) = 0,$

which is the required equation whose roots are $(\beta + \gamma), (\gamma + \alpha), (\alpha + \beta)$.

Now from the equation (3), we have

$$\Sigma(\beta + \gamma)(\gamma + \alpha) = (-1)^2 \cdot \frac{\text{Coefficient of } y}{\text{Coefficient of } y^3} = \frac{p^2 + q}{1} = p^2 + q$$

and $(\beta + \gamma)(\gamma + \alpha)(\alpha + \gamma) = (-1)^3 \cdot \dfrac{\text{Constant term}}{\text{Coefficient of } y^3}$

$$= -\frac{(pq - r)}{1} = r - pq.$$

Example 17:

If the roots of the equation $x^3 - 6x^2 + 11x - 6 = 0$ be α, β, γ, find the equation whose roots are $\beta^2 + \gamma^2, \gamma^2 + \alpha^2, \alpha^2 + \beta^2$.

Solution:

The given equation is

$$x^3 - 6x^2 + 11x - 6 = 0,$$

whose roots are α, β, γ.

$\therefore \Sigma\alpha = 6, \Sigma\alpha\beta = 11$ and $\alpha\beta\gamma = 6$.

Let y be a root of the required equation. Then

$$y = \beta^2 + \gamma^2 = (\Sigma\alpha^2) - \alpha^2 = (\Sigma\alpha)^2 - 2\Sigma\alpha\beta - \alpha^2 = 6^2 - 2 \times 11 - \alpha^2$$

or $y = 14 - \alpha^2$.

Replacing α be x which is a root of (1), we have

$$y = 14 - x^2 \text{ or } x^2 = 14 - y.$$

The required equation is obtained by eliminating x between (1) and (2) for which we proceed as follows.

The equation (1) can be written as $x(x^2 + 11) = 6(x^2 + 1)$.

Squaring both sides, we get $x^2(x^2 + 11)^2 = 36(x^2 + 1)^2$.

Substituting $x^2 = 14 - y$ from (2) in (3), we get

$(14 - y)(14 - y + 11)^2 = 36(14 - y + 1)^2$

or $(14 - y)(25 - y)^2 = 36(15 - y)^2$ or $y^3 - 28y^2 + 245y - 650 = 0$,

which is the required equation.

Example 18:

If α, β, γ are the roots of the equation $x^3 + qx + r = 0$ find the equations whose roots are

(i) $\frac{\beta}{\gamma}+\frac{\gamma}{\beta}, \frac{\gamma}{\alpha}+\frac{\alpha}{\gamma}, \frac{\alpha}{\beta}+\frac{\beta}{\alpha}$, *and*

(ii) $\frac{\beta\gamma}{\alpha}, \frac{\gamma\alpha}{\beta}, \frac{\alpha\beta}{\gamma}$.

Solution:

The given equation is $x^3 + qx + r = 0$,

whose roots are α, β, γ.

$\therefore\ \Sigma\alpha = 0$, $\Sigma\alpha\beta = q$ and $\alpha\beta\gamma = -r$.

(i) Let y be a root of the required equation. Then

$$y = \frac{\beta}{\gamma}+\frac{\gamma}{\beta} = \frac{\beta^2+\gamma^2}{\beta\gamma} = \frac{(\beta+\gamma)^2 - 2\beta\gamma}{\beta\gamma} = \frac{(\Sigma\alpha-\alpha)^2 - 2\beta\gamma}{\beta\gamma}$$

$$= \frac{(0-\alpha)^2 - 2\beta\gamma}{\beta\gamma} = \frac{\alpha^2}{\beta\gamma} - 2 = \frac{\alpha^3}{\alpha\beta\gamma} - 2 = \frac{\alpha^3}{-r} - 2 \quad [\because\ \alpha\beta\gamma = -1]$$

or $y = -\frac{x^3}{r} - 2$ [Replacing α by x which is a root of (1)]

or $x^3 + r(y + 2) = 0$.

The required equation is obtained by eliminating x between (1) and (2) for which we proceed as follows.

Subtracting (2) from (1), we get

$$qx - ry - r = 0 \text{ or } x = \frac{r(y+1)}{q}.$$

Putting this value of x in (2), we get

$$\frac{r^3}{q^3}(y+1)^3 + r(y+2) = 0 \text{ or } r^2(y^3 + 3y^2 + 3y + 1) + q^3(y + 2) = 0$$

or $r^2y^3 + 3r^2y^2 + (3r^2 + q^3)y + (r^2 + 2q^3) = 0$,

which is the required equation.

(ii) If y is a root of the required equation, then

$$y = \frac{\beta\gamma}{\alpha} = \frac{\alpha\beta\gamma}{\alpha^2} = \frac{-r}{\alpha^2} \qquad [\because \alpha\beta\gamma = -r]$$

or $\quad y = -\frac{r}{x^2}$ [Replacing α by x which is a root of (1)]

or $\quad x^2 = -\frac{r}{y}$.

The required equation is obtained by eliminating x between (1) and (3) for which we proceed as follows:

From (1), we have

$x(x^2 + q) + r = 0$ or $x\left(-\frac{r}{y} + q\right) + r = 0$ [Substituting from (3)]

or $\quad x = \frac{ry}{r - qy}$.

Putting this value of x in (3), we get

$$\left(\frac{ry}{r - qy}\right)^2 = -\frac{r}{y}$$

or $\quad r^2y^3 = -r(r - qy)^2$

or $\quad ry^3 + q^2y^2 - 2rqy + r^2 = 0,$

which is the required equation.

Example 19:

The roots of the equation $x^3 + px^2 + qx + r = 0$ are α, β, γ from an equation whose roots are

$\beta^2 + \gamma^2, \gamma^2 + \alpha^2, \alpha^2 + \beta^2$.

Solution:

Let us first form an equation whose roots are $\alpha^2, \beta^2, \gamma^2$.

The given equation can be written as $px^2 + r = -x(x^2 + q)$.

Squaring both sides, we get

$(px^2 + r)^2 = x^2(x^2 + q)^2$. ...(1)

Replacing x^2 by y in (1), we get

$(py + r)^2 = y(y + q)^2$

or $y^3 - (p^2 - 2q)y^2 + (q^2 - 2pr)y - r^2 = 0$. ...(2)

Thus, $\alpha^2, \beta^2, \gamma^2$ are the roots of the equation (2).

Let $\alpha^2 = A, \beta^2 = B, \gamma^2 = C$.

Now if A, B, C be the roots of the equation (2), to form an equation whose roots are

$B + C, C + A, A + B$.

Let $z = B + C = (A + B + C) - A = p^2 - 2q - A$.

$$[\because \Sigma A = p^2 - 2q]$$

$\therefore A = p^2 - 2q - z = k - z$, where $k = p^2 - 2q$.

But A is a root of the equation (2). Therefore the required transformed equation is

$$(k - z)^3 - k(k - z)^2 + (q^2 - 2pr)(k - z) - r^2 = 0$$

or $k^3 - 3k^2z + 3kz^2 - z^3 - k^3 + 2k^2z - kz^2 + k(q^2 - 2pr) - (q^2 - 2pr)$
$- (q^2 - 2pr)z - r^2 = 0$

or $z^3 - 2kz^2 + (k^2 + q^2 - 2pr)z - k(q^2 - 2pr) + r^2 = 0$

or $z^3 - 2(p^2 - 2q)z^2 + (p^4 - 4p^2q + 5q^2 - 2pr)z - (p^2q^2 - 2p^3r + 4pqr$
$s - 2p^3 - r^2) = 0$.

Example 20:

If α, β, γ be the roots of the equation $2x^3 + x^2 + x + 1 = 0$, find the equation whose roots are

$$\frac{1}{\beta^2} + \frac{1}{\gamma^2} - \frac{1}{\alpha^2}, \frac{1}{\gamma^2} + \frac{1}{\alpha^2} - \frac{1}{\beta^2}, \frac{1}{\alpha^2} + \frac{1}{\beta^2} - \frac{1}{\gamma^2}.$$

Solution:

The given equation is $2x^3 + x^2 + x + 1 = 0$. ...(1)

Let us first find the equation whose roots are the squares of the roots of (1). We may put (1) in the form

$$x^2 + 1 = -x(2x^2 + 1).$$

Squaring both sides, we get $(x^2 + 1)^2 = x^2(2x^2 + 1)^2$.

Put $x^2 = y$. Then $\alpha^2, \beta^2, \gamma^2$ are the roots of the equation

$$(y + 1)^2 + y(2y + 1)^2$$

or $4y^3 + 3y^2 - y - 1 = 0$. ...(2)

$\therefore \frac{1}{\alpha^2}, \frac{1}{\beta^2}, \frac{1}{\gamma^2}$ are the roots of the equation

$$z^3 + z^2 - 3z - 4 = 0. \qquad ...(3)$$

Let A, B, C be the roots of the equation (3). Then we are to find the equation whose roots are

$$B + C - A,\ C + A - B,\ A + B - C.$$

Let $\quad t = B + C - A = (A + B + C) - 2A = -1 - 2A.$

$\therefore \quad A = -\dfrac{t+1}{2}.$

But A is a root of the equation (3). Hence the transformed equation is

$$\left(-\frac{t+1}{2}\right)^3 + \left(-\frac{t+1}{2}\right)^2 + 3\left(\frac{t+1}{2}\right) - 4 = 0$$

or $\quad t^3 - t^2 - 13t + 19 = 0.$

Example 21:

If α, β, γ be the roots of the cubic $ax^3 + 3bx^2 + 3cx + d = 0$ prove that the equation in y whose roots are

$$\frac{\beta\gamma - \alpha^2}{\beta + \gamma - 2\alpha}, \frac{\gamma\alpha - \beta^2}{\gamma + \alpha - 2\beta}, \frac{\alpha\beta - \gamma^2}{\alpha + \beta - 2\gamma}$$

is obtained by the transformation $axy + b(x + y) + c = 0$.

Hence or otherwise form the equation with the above roots.

Solution:

Let

$$y = \frac{\beta\gamma - \alpha^2}{\beta + \gamma - 2\alpha} = \frac{\alpha\beta\gamma - \alpha^3}{\alpha(\alpha + \beta + \gamma - 3\alpha)} = \frac{\alpha\beta\gamma - \alpha^3}{\alpha[(\Sigma\alpha) - 3\alpha]} = \frac{-\frac{d}{d} - \alpha^3}{\alpha\left[-\frac{3b}{a} - 3\alpha\right]}.$$

Replacing α by x, is any root of the given cubic, we get

$$y = \frac{-\frac{d}{a} - x^3}{x\left[-\frac{3b}{a} - 3x\right]}$$

or $\quad xy\left(\dfrac{3b}{a} + 3x\right) = \dfrac{d}{a} + a + x^3$

or $\quad 3xyb + 3ax^2y = d + ax^3$

or $\quad ax^3 - 3ayx^2 - 3byx + d = 0.$ ...(1)

Also $\quad ax^3 + 3bx^2 + 3cx + d = 0.$ ...(2)

Subtracting (1) from (2), we get

$$3x^2(b + ay) + 3x(c + by) = 0$$

i.e., $x(bx + axy + c + by) = 0$ *i.e.* $bx + by + axy + c = 0$, since $x \neq 0$.

From this, we get

$$x = -\frac{c + by}{b + ay}.$$

Putting this value of x in (2), we get the required equation as

$$(2ab^3 - 3a^2bc + a^3d)\, y^3 + (3b^4 - 3ab^2c + 3a^2bd - 3a^2c^2)y^2 + (3b^3c + 3ab^2d - 6abc^2)\, y + b^3d - ac^3 = 0.$$

Example 22:

Change the signs of the roots of the equation

$x^5 - 4x^3 + 3x^2 + 8x - 9 = 0.$

Solution:

Supplying the missing terms with zero coefficients, the given equation after making it complete may be written as

$$x^5 + 0x^4 - 4x^3 + 3x^2 + 8x - 9 = 0. \quad ...(1)$$

Changing the sign of every alternate term of the equation (1) beginning with the second, the required transformed equation is

$$x^5 - 0x^4 - 4x^3 - 3x^2 + 8x + 9 = 0$$

or $x^5 - 4x^3 - 3x^2 + 8x + 9 = 0.$

EXERCISES

1. Find the equations whose roots are the cubes of the roots of the following equations:

(i) $x^3 + ax^2 + bx + ab = 0.$

(ii) $x^4 - x^3 + 2x^2 + 3x + 1 = 0.$

(iii) $x^3 + 3x^2 + 2 = 0.$

2. Find the equation whose roots are the squares of the roots of the equation

$x^3 + px^2 + qx + r = 0$

and hence find the value of $\Sigma\alpha^2\beta^2$ for the given cubic, α, β, γ being its roots.

3. Find the equation whose roots are the squares of the roots of the roots of

   $x^4 + bx^2 + cx + d = 0.$

4. Find the equation whose roots are the squares of the roots of the quadratic equation $x^4 + bx^2 + cx + d = 0$.

5. Solve the following reciprocal equations:

(i) $x^4 - 10x^3 + 26x^2 - 10x + 1 = 0.$

(ii) $6x^6 - 25x^5 + 31x^4 - 31x^2 + 25x - 6 = 0.$

6. The roots of the equation $81x^3 - 18x^2 - 36x + 8 = 0$ are in harmonic progression. Transform it into another with integral coefficients and unity for the coefficient of the first term, so that the roots of the transformed equation may be in arithmetical progression. Hence, solve the equation.

7. Remove the fractional coefficients from the equation

   $$x^4 - \frac{5}{6}x^3 - \frac{13}{12}x^2 + \frac{1}{300} = 0.$$

8. Remove the fractional coefficients from the equation

   $$x^3 - \frac{5}{2}x^2 - \frac{7}{18}x + \frac{1}{108} = 0.$$

9. Find the equation whose roots are twice the reciprocals of the roots of

   $x^4 + 3x^3 - 6x^2 + 2x - 4 = 0.$

10. Form the equation whose roots are the reciprocals of the roots of the equation

    $x^4 - 3x^3 + 7x^2 + 5x - 2 = 0.$

11. Transform the equation $5x^3 - \frac{3}{2}x^2 - \frac{3}{4}x + 1 = 0$ into another with integral coefficients and unity for the coefficient of the leading term.

12. Change the signs of the roots of the equation $x^7 + 5x^5 - x^3 + x^2 + 7x + 3 = 0$.

13. Find the condition that the 2nd and 3rd terms of the cubic

    $a_0x^3 + 3a_1x^2 + 3a_2x + a_3 = 0,$

    are removed by the same transformation.

14. Compute H and G for the cubic $x^3 + 6x^2 + 12x - 19 = 0$.

15. Compute H and G for the cubic $x^3 + 3x^2 + 4x - 10 = 0$.

16. Transform the equation $x^4 - 4x^3 - 18x^2 - 3x + 2 = 0$, into one which shall want the third term.

17. Transform the equation $x^4 - 8x^3 + x^2 - x + 3 = 0$ into an equation lacking the second term.

18. Transform each of the following equations into an equation lacking the second term:

(i) $x^4 + 8x^3 + x - 5 = 0$.

(ii) $x^4 + 20x^3 + 143x^2 + 430x + 462 = 0$.

19. Show that the same transformation removes both 2nd and 4th terms of the equation

$x^4 + 16x^3 + 83x^2 + 152x + 84 = 0$, and find its roots.

20. Show that the roots of the equation $x^3 - \frac{1}{3}q^2x - \frac{2}{27}q^3 - r^2 = 0$, differ by a constant from the squares of the corresponding roots of $x^3 + qx + r = 0$.

21. If α, β, γ are the roots of $8x^3 - 4x^2 + 6x - 1 = 0$ find the equation whose roots are

$\alpha + \frac{1}{2}, \beta + \frac{1}{2}, \gamma + \frac{1}{2}$.

22. Find the equation whose roots are those of $3x^3 - 2x^2 + x - 9 = 0$, each diminished by 5.

23. Find the equation whose roots are the roots of the equation $4x^5 - 2x^3 + 7x - 3 = 0$, each increased by 2.

24. Find the equation each of the whose roots is greater by unity than a root of

$x^3 - 5x^2 + 6x - 3 = 0$.

25. Find the equation whose roots are equal to the roots of

$x^4 + x^3 - 3x^2 - x + 2 = 0$,

each diminished by 3.

26. If the roots of the cubic $x^3 + 2x^2 + 3x + 1 = 0$ be α, β, γ, find equation whose roots are

$$\frac{1}{\beta^3}+\frac{1}{\gamma^3}-\frac{1}{\alpha^3}, \frac{1}{\gamma^3}+\frac{1}{\alpha^3}-\frac{1}{\beta^3}, \frac{1}{\alpha^3}+\frac{1}{\beta^3}-\frac{1}{\gamma^3}.$$

27. If α, β, γ are the roots of the equation $x^3 + qx + r = 0$, form the equation whose roots are

(i) $\beta^2\gamma^2, \gamma^2\alpha^2, \alpha^2\beta^2$

(ii) $\beta^2 + \beta\gamma + \gamma^2, \gamma^2 + \gamma\alpha + \alpha^2, \alpha^2 + \alpha\beta + \beta^2$

(iii) $\alpha^2(\beta + \gamma), \beta^2(\gamma + \alpha), \gamma^2(\alpha + \beta)$.

28. If α, β, γ are the roots of the equation $x^3 + px^2 + qx + r = 0$, find the equation whose roots are $\alpha^2 + 2\beta\gamma, \beta^2 + 2\gamma\alpha, \gamma^2 + 2\alpha\beta$.

29. If α, β, γ are the roots of the cubic $x^3 + qx + r = 0$ find the equation whose roots are

$(\alpha - \beta)^2, (\beta - \gamma)^2, (\gamma - \alpha)^2$.

30. If α, β, γ are the roots of the equation $x^3 + qx + r = 0$, find the equations whose roots are

(i) $\frac{1}{\beta}+\frac{1}{\gamma}, \frac{1}{\gamma}+\frac{1}{\alpha}, \frac{1}{\alpha}+\frac{1}{\beta}$

(ii) $\frac{\beta+\gamma}{\alpha^2}, \frac{\gamma+\alpha}{\beta^2}, \frac{\alpha+\beta}{\gamma^2}$

(iii) $\frac{\beta^2+\gamma^2}{\alpha}, \frac{\gamma^2+\alpha^2}{\beta}, \frac{\alpha^2+\beta^2}{\gamma}$

(iv) $\frac{\beta^2+\gamma^2}{\alpha^2}, \frac{\gamma^2+\alpha^2}{\beta^2}, \frac{\alpha^2+\beta^2}{\gamma^2}$ and

(v) $\alpha-\frac{1}{\beta\gamma}, \beta-\frac{1}{\alpha\gamma}, \gamma-\frac{1}{\alpha\beta}$. **(Meerut, 1992)**

31. If α, β, γ are the roots of the cubic $x^3 - px^2 + qx - r = 0$, find the equation whose roots are

$(\beta + \gamma - \alpha), (\gamma + \alpha - \beta), (\alpha + \beta - \gamma)$.

Also find the value of $(\beta + \gamma - \alpha)(\gamma + \alpha - \beta)(\alpha + \beta - \gamma)$.

32. If α, β, γ are the roots of the equation $x^3 + px^2 + qx + r = 0$, find the equations whose roots are

(i) $\beta\gamma + \frac{1}{\alpha}, \gamma\alpha + \frac{1}{\beta}, \alpha\beta + \frac{1}{\gamma}$

(ii) $\alpha - \frac{1}{\beta\gamma}, \beta - \frac{1}{\gamma\alpha}, \gamma - \frac{1}{\alpha\beta}$

(iii) $\alpha - \frac{\beta\gamma}{\alpha}, \beta - \frac{\gamma\alpha}{\beta}, \gamma - \frac{\alpha\beta}{\gamma}$

(iv) $\frac{\alpha}{\beta + \gamma}, \frac{\beta}{\gamma + \alpha}, \frac{\gamma}{\alpha + \beta}$ and

(v) $\alpha(\beta + \gamma), \beta(\gamma + \alpha), \gamma(\alpha + \beta)$.

3

DESCARTE'S RULE OF SIGNS

3.1 CHARACTER AND POSITION OF THE ROOTS OF AN EQUATION

The *character* of the roots of an equation is known if we find the exact number of real and imaginary roots separately. The *position* of a real root, when it is integral, is determined by finding its exact value; and when it is non-integral, it is determined by finding two consecutive integers between which the root lies.

Here in this chapter we shall give some theorems which help us in determining the character and position of the roots of an equation.

3.2 AN IMPORTANT PROPERTY OF A POLYNOMIAL

Theorem:

If two real numbers a and b be substituted for x in any polynomial f(x), and if f(a) and f(b) are of opposite signs, then at least one or an odd number of real roots of the equation f(x) = 0 lie between a and b. And if f(a) and f(b) be of the same sign, then either no real root or an even number of roots of f(x) = 0 lie between a and b. **(Agra, 1991)**

We know that a polynomial is a continuous function of x. Therefore, as x passes through all the values from a to b, f(x) must pass through all the values from f(a) to f(b). Hence if we put y = f(x), the graph of this function must be a continuous curve between x = a and x = b.

Case I: f(a) and f(b) of Opposite Signs: In this case the points of the values x = a and x = b lie on the opposite sides of the axis of x. Therefore while going from one side of x-axis to the other side of x-axis the curve y = f(x) must cross the axis of x at least once or an odd number of times. Thus, between x = a and x = b, there must lie at least one or an odd number of real values of x for which y = 0 *i.e.*, f(x) = 0. Hence in this case at least one or an odd number of roots of the equation f(x) = 0 lie between a and b.

Case II: f(a) and f(b) of the same sign: In this case the points of the curve y = f(x) corresponding to the values x = a and y = b lie on the same side of the axis of x. Therefore, while going from one point to the other point, either the curve y = f(x) must not cross the axis of x or it must cross it an even number of times. Therefore in this case either no real root or an even number of real roots of the equation f(x) = 0 lie between a and b.

Corollary: *If a polynomial f(x) keeps its sign constant for all real values of x, then the equation f(x) = 0 will have no real roots.*

3.3 DESCARTE'S RULE OF SIGNS

It is possible to determine the nature of the roots of an equation f(x) = 0 without actually determining its roots. One such method is given by Descrate called the Descarte's rule of signs. For this, we require the number of changes of signs that occur in a polynomial whose terms are arranged in an order (descending or ascending).

Examples:

Consider the polynomials

$f(x) = x^6 - 7x^4 + 8x^3 + 2x^2 - 9x - 5$

$g(x) = x^7 + 8x^6 - 2x^5 + 4x^3 - 7x^2 + 2$

$h(x) = x^4 + 2x^3 - 3x^2 - 5x - 1$

$\phi(x) = 6x^4 + 3x^2 + 7x + 9.$

In f(x), we have + – + + – + *i.e.*, 3 changes of signs.

In g(x), we have + + – + – + *i.e.*, 4 changes of signs.

In h(x) we have + + – – – *i.e.*, 1 change of sign.

In ϕ(x), we have + + + + *i.e.*, no change of sign.

Descarte's Rule of Signs

Positive Roots: *An equation f(x) = 0 cannot have more positive roots than the number of changes of signs from + to – and from – to + in the terms of its first member f(x).*

The rigorous proof of this theorem is beyond the scope of this book. Only a verification of the above statement is given here.

Let the signs of the terms taken in descending order in any polynomial chosen at random be

$+ + - - - + - +.$

In this polynomial there are in all four changes of signs. Let us multiply this polynomial by binomial x – h corresponding to a positive root h. The signs of this binomial are + – . Since we are concerned only with the signs, therefore the process of multiplication can be shown as below:

$$+ + - - - + - + \quad ...(1)$$

$$+ -$$

$$+ + - - - + - +$$

$$- - + + + - + -$$

$$+ \pm - \pm \pm + - + - \quad ...(2)$$

Now, we proceed to show that the resulting polynomial (2) has at least one more change of signs than the original polynomial (1). There are three ambiguous signs in the resulting polynomial. Let u first take the most unfavourable case in which the continuation of sign in the original polynomial remains continuation in the resulting polynomial. The signs of the resulting polynomial in this case are

$$+ + - - - + - + -.$$

This gives 5 changes of signs, the series of signs being the same as the original one in (1) with an additional change of signs at the end.

Thus, the multiplication of a polynomial by x – h, where h is positive, increases the number of changes of signs by at least one.

Suppose now $\phi(x)$ is a polynomial formed of the products of the factors corresponding to the negative and imaginary roots of an equation f(x) = 0. Then, if α, β, γ, ... are the real positive roots of f(x) = 0, successive multiplications by $x - \alpha$, $x - \beta$, ... will each increase the number of changes of signs in $\phi(x)$ by at least one. Hence, in the complete equation there will be at least as many changes of signs as it has positive roots. Thus, we can say that an equation cannot have more positive roots than it has changes of signs.

Corollary 1: Negative Roots: *The equation f(x) = 0 cannot have more negative roots than the number of changes of signs in f(–x).*

We know that the negative roots of the equation f(x) = 0 are the positive roots of the equation f(–x) = 0. Therefore by the Descarte's rule of signs for positive roots, it follows that the number of the negative roots of the equation f(x) = 0 cannot exceed the number of changes of signs in f(–x).

Corollary 2: Complex Roots: Suppose for a given equation f(x) = 0 of the nth degree the number of changes of signs in f(x) is a and the number of changes of signs in $f(-x)$ is b. Then the given equation cannot have more than $a + b$ real roots. Therefore if $a + b$ is less than n, we can conclude that the given equation must have at least $n - (a + b)$ complex roots. This will happen only when the given equation is *incomplete.*

In case the polynomial equation $f(x) = 0$ is of degree n and is *complete,* the number of changes of signs in $f(x)$ *i.e.,* the maximum possible number of positive roots and the number of changes of signs in $f(-x)$ *i.e.,* the maximum possible number of negative roots taken together is n *i.e.,* the degree of the equation and as such we cannot draw any definite conclusion regarding the existence of imaginary roots. The conclusion regarding the existence of complex roots can only be drawn when the given equation is incomplete.

For example, consider the equation $f(x) \equiv x^7 - 3x^5 + 9x^4 + 2x^3 - 5 = 0$.

The above equation has three changes of signs and as such it cannot have more than three positive roots. Again the equation $f(-x) = 0$

i.e., $\quad -x^7 + 3x^5 + 9x^4 - 2x^3 - 5 = 0$

i.e., $\quad x^7 - 3x^5 - 9x^4 + 2x^3 + 5 = 0.$

has only two changes of signs and as such $f(x) = 0$ cannot have more than two negative roots. Also, 0 is not a root of the given equation. Hence, the maximum number of real roots of the given equation is 3 + 2 *i.e.,* 5 and the degree of the equation being 7 we conclude that the equation must have at least two imaginary roots.

3.4 SOME IMPORTANT DEDUCTIONS FROM DESCARTE'S RULE OF SIGNS

Deduction 1:

Every equation of an odd degree has at least one real root whose sign is opposite to that of its last term, the coefficient of the first term being positive.

Proof:

Let the equation be

$$f(x) \equiv x^n + p_1x^{n-1} + p_2x^{n-2} + \dots + p_n = 0, \text{ (n is odd).}$$

We have $f(-\infty) = -$ ve (n being odd),

$f(0) = p_n$, $f(\infty) = +$ ve.

Case I: p_n is positive: In this case $f(-\infty)$ and $f(0)$ are of opposite signs and hence at least one real root must lie between $-\infty$ and 0 *i.e.,* it must be negative opposite to the sign of p_n which is + ve.

Case II: p_n is negative: In this case

$$f(0) = p_n = -\text{ ve and } f(\infty) = +\text{ ve}$$

i.e., they are of opposite signs and hence at least one real root must lie between 0 and ∞. This root is clearly +ve *i.e.*, of the sign opposite to that of p_n which is – ve.

Deduction 2:

Every equation of even degree whose last term is negative and the coefficient of the first term positive, has at least two real roots, one positive and one negative.

Proof:

Let the equation be

$$f(x) \equiv x^n + p_1x^{n-1} + p_2x^{n-2} + \ldots + p_n = 0, \text{ (n is even)}.$$

We have $f(-\infty) = +$ ve, $f(0) = p_n$

i.e., – ve and $f(\infty) = +$ ve.

Here $f(-\infty)$ and $f(0)$ are of opposite signs and so at least one real root must lie between $-\infty$ and 0 and it is negative. Again $f(0)$ and $f(\infty)$ are of opposite signs and hence at least one real root must lie between 0 and ∞ and it is positive. Thus, the equation must have at least real roots, one + ve and the other – ve.

Deduction 3:

If an equation has only one change of signs, it must have only one positive root and no more.

Proof:

Taking the leading coefficient to be + ve, the equation $f(x) = 0$ must have a set of + ve terms followed by a set of – ve terms since there is only one change of signs. Consequently the last term in the equation *i.e.*, the constant term is negative.

We have $f(\infty) = +$ ve and $f(0) =$ the last term which is – ve.

Thus, $f(0)$ and $f(\infty)$ are of opposite signs and as such there must lie at least one or an odd number of roots of the equation $f(x) = 0$ between 0 and ∞. But as there is only change of signs in $f(x)$, the number of positive roots of $f(x) = 0$ cannot be greater than one. Hence, the equation $f(x) = 0$ must have one and only one positive root.

Deduction 4:

If all the terms of an equation are positive and the equation involves no odd powers of x, then all its roots are complex.

Proof:

Suppose all the terms of the equation $f(x) = 0$ are positive and the equation has no odd powers of x. Clearly both $f(x)$ and $f(-x)$ will have no changes of signs and hence by Descarte's rule of signs there will not be any + ve or – ve roots of $f(x) = 0$. Hence, all the roots of $f(x) = 0$ must be complex.

Deduction 5:

If all the terms of an equation are positive and the equation involves only odd powers of x, then 0 is the only real root.

Proof:

Let all the terms of the equation $f(x) = 0$ be positive and let $f(x)$ involve only odd powers of x.

Obviously, the equation $f(x) = 0$ does not contain the contain the constant term and so $x = 0$ must be a root of the equation.

Now, if we divide the equation $f(x) = 0$ by x, then the resulting equation $g(x) = 0$ contains only positive terms and $g(x)$ contains only even powers of x. Hence, all the roots of $g(x) = 0$ must be complex. Consequently 0 is the only real root of $f(x) = 0$.

SOLVED EXAMPLES

Example 1:

Find the equation whose roots are the squares of the roots of the equation $x^3 + 4x^2 + 9x + 10 = 0$ and hence find the nature of the roots of the given equation.

Solution:

The given equation is

$$x^3 + 4x^2 + 9x + 10 = 0$$

or $$x(x^2 + 9) = -(4x^2 + 10).$$

Squaring both sides, we get

$$x^2(x^2 + 9)^2 = (4x^2 + 10)^2.$$

Replacing x^2 by y, the required equation whose roots are the squares of the roots of the given equation is

$$y(y + 9)^2 = (4y + 10)^2 \text{ or } f(y) \equiv y^3 + 2y^2 + y - 100 = 0.$$

The number of changes of signs in f(y) is 1 and so the equation f(y) = 0 cannot have more than one positive root. We have $f(\infty) = +$ ve and $f(0) = -$ ve. So, the equation f(y) = 0 has one and only one positive and two imaginary root.

Since the roots of f(y) = 0 are the squares of the roots of $x^3 + 4x^2 + 9x + 10 = 0$, therefore the roots of f(y) = 0 cannot be negative. So, the roots of f(y) = 0 are either all positive or one positive and two imaginary.

So, the equation f(y) = 0 has one positive and two imaginary roots. The two imaginary roots of f(y) = 0 correspond to the two imaginary roots of $x^3 + 4x^2 + 9x + 10 = 0$.

Hence, the equation $x^3 + 4x^2 + 9x + 10 = 0$ must have a pair of imaginary roots.

Since the equation $x^3 + 4x^2 + 9x + 10 = 0$ is of odd degree, therefore it has at least one real root which cannot be positive because $x^3 + 4x^2 + 9x + 10 = 0$ has no change of sign.

Hence, the equation $x^3 + 4x^2 + 9x + 10 = 0$ has one negative and two complex roots.

Example 2:

Locate the situation of the roots of the equation $x^3 + x^2 - 2x - 1 = 0$.

Solution:

The given equation is

$$f(x) \equiv x^3 + x^2 - 2x - 1 = 0.$$

The equation f(– x) = 0 is

$$-x^3 + x^2 + 2x - 1 = 0 \text{ i.e., } x^3 - x^2 - 2x + 1 = 0.$$

We see that f(x) has only one change of sign and so the equation f(x) = 0 cannot have more than one positive root.

We have $f(0) = -1$, $f(1) = -1$, $f(2) = 7$.

Since f(1) and f(2) are of opposite signs, therefore the positive root lies between 1 and 2.

Again f(–x) and f(0) are of opposite signs, therefore one negative root lies between –1 and 0. Again f(–2) and f(–1) are also of opposite signs and so one negative root lies between –2 and –1.

Hence, all the three roots of the given equation are real and they lie in the open intervals (–2, –1), (–1, 0) and (1, 2).

Example 3:

Apply Descarte's Rule of signs to discuss the nature of the roots of the equation

$$x^4 + 15x^2 + 7x - 11 = 0.$$ **(Agra, 1994)**

Solution:

The given equation is

$$f(x) \equiv x^4 + 15x^2 + 7x - 11 = 0.$$

The number of changes of signs in f(x) is 1. So by Descarte's rule of signs, the equation f(x) = 0 cannot have more than one positive root.

We have $f(\infty) = +$ ve and $f(0) = -11 = -$ ve.

Since f(0) and $f(\infty)$ are of opposite signs, therefore the equation f(x) = 0 has at least one or an odd number of roots lying between 0 and ∞.

Hence, the equation f(x) = 0 has one and only one positive root.

Again $f(-x) \equiv x^4 + 15x^2 - 7x - 11$.

The number of changes of signs in f(–x) is 1 and so the equation f(–x) = 0 cannot have more than one positive root. Consequently the equation f(x) = 0 cannot have more than one negative root.

We have $f(-\infty) = +$ ve and $f(0) = -11 = -$ ve.

Since $f(-\infty)$ and f(0) are of opposite signs, therefore the equation f(x) = 0 has at least one or an odd number of roots lying between $-\infty$ and 0.

Thus, f(x) = 0 has two real roots, one + ve and one – ve and hence the other two roots must be imaginary.

Example 4:

Show that the equation $x^5 + x^3 - 8x - 5 = 0$ cannot have more than three real roots and prove that it must have three real roots.

Solution:

Let $f(x) \equiv -x^5 - x^3 + 8x - 5 = 0$

or $x^5 + x^3 - 8x + 5 = 0.$

Since f(– x) = 0 has only two changes of signs, therefore the equation f(x) = 0 cannot have more than 2 negative roots. Thus the equation f(x) = 0 cannot have more than three real roots.

Now $f(0) = -5 = -$ ve and $f(\infty) = +\infty = +$ ve.

Since f(0) and f(∞) are of opposite signs, therefore at least one or an odd number of real roots of the equation f(x) = 0 lie between 0 and ∞.

But, we have already shown that this equation cannot have more than one positive root. Hence it must have one and only one positive root.

Again f(0) = – ve, f(– 1) = + ve and f(– ∞) = – ve.

Now, the equation f(x) = 0 cannot have more than two negative roots. Since f(0) and f(– 1) are of opposite signs and f(–1) and f(– ∞) are also of opposite signs, therefore the equation f(x) = 0 has one root lying between – 1 and 0 and one between – ∞ and – 1.

Hence, the given equation cannot have more than three real roots and it must have three real roots one positive and two negative.

Example 5:

Show that the equation $2x^7 - x^4 + 4x^3 - 5 = 0$ *has at least four imaginary roots.*

Solution:

We see that f(x) has only three changes of signs. Therefore the equation f(x) = 0 cannot have more than three positive roots.

Again $\quad f(-x) \equiv -2x^7 - x^4 - 4x^3 - 5 = 0,$

or $\quad 2x^7 + x^4 + 4x^3 + 5 = 0.$

We see that f(–x) has no change of signs. Therefore the equation f(x) = 0 cannot have any negative root. Thus the maximum number of real roots of the equation f(x) = 0 is 3. But this equation is of degree 7. Hence it must have at least 7 – 3 *i.e.,* 4 complex roots.

EXERCISES

1. Show that for all real values of k the equation $2x^5 + 3x^2 + 5c + k = 0$ has at least two imaginary roots.
2. Show that the equation $x^n + 1 = 0$ has, when n is even, no real root; and when n is odd, the real root – 1, and no other real root.
3. Show that the equation $x^n - 1 = 0$ has, when n is even, two real roots 1 and – 1, and no other real root; and when n is odd, the real root 1, and no other real root.
4. Show that the equation $x^4 - ax^3 - bx - c = 0$ has one positive root, one negative root and two complex roots for all positive values of a, b and c.

5. Show that the equation $x^3 - qx + r = 0$ where q and r are essentially positive has one negative root and that the other two roots are either imaginary or both positive.
6. Prove that the equation $x^5 - x + 16 = 0$ has two pairs of complex roots.
7. Show that the equation

   $$x^9 + x^5 + x^4 + x^2 + 1 = 0$$

   has only one real root which is negative.
8. Prove that the equation

   $$4x^3 - 13x^2 - 31x - 275 = 0$$

   has one positive root and show that it lies between 6 and 7.
9. Find the minimum number of imaginary roots which the equation

   $$2x^7 - x^4 + 4x^3 - 5 = 0$$

   must posses.
10. Show that the equation $x^{12} - x^4 + x^3 - x^2 + 1 = 0$ has at least six complex roots.
11. Show that the equation $x^9 - x^5 + x^4 + x^2 + 1 = 0$ has at least six complex roots.
12. Show that the equation

    $$x^6 + 3x^2 - 5x + 1 = 0$$

    has at least four imaginary roots.

4

SOLUTION OF CUBIC EQUATIONS

4.1 CARDAN'S METHOD OF SOLVING THE CUBIC EQUATIONS

(Meerut, 1988, 89, 90)

Let the cubic equation be $ax^3 + 3bx^2 + 3cx + d = 0$. ...(1)

Removing the second term and multiplying the roots by a, the equation (1) can be reduced to the form

$$z^3 + 3Hz + G = 0 \qquad ...(2)$$

where $z = a\left(x + \frac{b}{a}\right) = ax + b$, G and H have their usual meanings

i.e., $G = a^2d - 3abc + 2b^3$ and $H = ac - b^2$.

Let us assume that $z = u^{1/3} + v^{1/3}$. ...(3)

i.e., $z^3 - 3u^{1/3}\,v^{1/3}z - (u + v) = 0$. ...(4)

Comparing the coefficients in (2) and (4), we get

$$u^{1/3}v^{1/3} = -H \text{ or } uv = -H^3$$

and $(u + v) = -G$.

Hence u and v are the roots of the quadratic $t^2 + Gt - H^3 = 0$. ...(5)

Solving (5), we get

$$u = \frac{-G + \sqrt{(G^2 + 4H^3)}}{2},\ v = \frac{-G - \sqrt{(G^2 + 4H^3)}}{2} \qquad ...(6)$$

Now by taking cube root we shall get three values of the cube root of u, namely, $u^{1/3}$, $\omega u^{1/3}$ and $\omega^2 u^{1/3}$. Similarly we shall get three values of the cube root of v, namely $v^{1/3}$, $\omega v^{1/3}$ and $\omega^2 v^{1/3}$. If we take all possible combinations, we shall get nine values of the expression $u^{1/3} + v^{1/3}$ which is a root of the equation (2). But a cubic should have only three roots. Therefore while combining the values of $u^{1/3}$ and $v^{1/3}$ we should not forget that they are also connected by the relation $u^{1/3}v^{1/3} = -H$. From this relation we get $v^{1/3} = \frac{-H}{u^{1/3}}$. Putting the value of $v^{1/3}$ in (3), we get $z = u^{1/3} - \frac{H}{u^{1/3}}$, the other two values of z being obtained by replacing $u^{1/3}$ by $\omega u^{1/3}$ and $\omega^2 u^{1/3}$ respectively.

Thus, if z_1, z_2, z_3 are the roots of the cubic (2), we have

$$z_1 = u^{1/3} - \frac{H}{u^{1/3}} = u^{1/3} + v^{1/3},$$

$$z_2 = \omega u^{1/3} - \frac{H}{\omega u^{1/3}} = \omega u^{1/3} - \omega^2 \frac{H}{u^{1/3}} = \omega u^{1/3} + \omega^2 v^{1/3},$$

and
$$z_3 = \omega^2 u^{1/3} - \frac{H}{\omega^2 u^{1/3}} = \omega^2 u^{1/3} - \omega \frac{H}{u^{1/3}} = \omega^2 u^{1/3} + \omega v^{1/3},$$

where $\omega = \frac{1}{2}\left(-1+\sqrt{3}i\right)$ and $\omega^2 = \frac{1}{2}\left(-1-\sqrt{3}i\right)$.

Having found the values of z, we can find the values of x by the relation $z = ax + b$.

4.2 APPLICATION OF CARDAN'S METHOD TO NUMERICAL EQUATIONS

When $G^2 + 4H^3 > 0$ or $= 0$ *i.e.,* when the cubic has two imaginary or two equal roots, the values of u and v as found from the quadratic in t are real. Now by some suitable arithmetical process we can extract the cube roots of real quantities and thus we shall get the values of $u^{1/3}$ and $v^{1/3}$. But if $G^2 + 4H^3 < 0$ *i.e.,* if all the roots of the cubic are real and different, the values of u and v are not real. Now, there is no general arithmetical process for extracting the cube root of complex numbers and consequently Cardan's method fails to give the solution of a cubic all of whose roots are real and different. This was called by older mathematicians *irreducible case of Cardan's solution.* However, in this case we can make use of De Moivre's Theorem of Trigonometry to find the values of the cube roots of complex numbers.

Let $-G = A$, $G^2 + 4H^3 = -B^2$,

so that $u = \frac{A}{2} + \frac{i}{2}B$ and $V = \frac{A}{2} - \frac{i}{2}B$.

$$\therefore\ z=\left(\frac{A}{2}+\frac{i}{2}B\right)^{1/3}+\left(\frac{A}{2}-\frac{i}{2}B\right)^{1/3}.$$

Now put $\frac{A}{2}=r\cos\theta$ and $\frac{B}{2}=r\sin\theta$, so that

$$r^2=\frac{A^2+B^2}{4}=\frac{G^2-G^2-4H^3}{4}=-H^3$$

and $\tan\theta=\frac{B}{A}=\frac{-\sqrt{\left[-\left(G^2+4H^3\right)\right]}}{G}.$

$$\therefore\ z=(r\cos\theta+ir\sin\theta)^{1/3}+(r\cos\theta-ir\sin\theta)^{1/3}$$

$$=r^{1/3}\left[\cos\left(\frac{2n\pi+\theta}{3}\right)+i\sin\left(\frac{2n\pi+\theta}{3}\right)\right.$$
$$\left.+\cos\left(\frac{2n\pi+\theta}{3}\right)-i\sin\left(\frac{2n\pi+\theta}{3}\right)\right]$$

$$=2r^{1/3}\cos\frac{2n\pi+\theta}{3},\ \text{where } n=0,1,2.$$

Hence the roots of the equation in z are $2r^{1/3}\cos\frac{\theta}{3}$, $2r^{1/3}\cos\frac{2\pi+\theta}{3}$ and $2r^{1/3}\cos\frac{4\pi+\theta}{3}$;

i.e., $2r^{1/3}\cos\frac{\theta}{3}$, $2r^{1/3}\cos\frac{2\pi+\theta}{3}$ and $2r^{1/3}\cos\frac{2\pi-\theta}{3}$;

i.e., $2(-H)^{1/2}\cos\frac{\theta}{3}$, $2(-H)^{1/2}\cos\frac{2\pi\pm\theta}{3}$.

3.3 EXPRESSING THE CUBIC AS SUM OR DIFFERENCE OF TWO CUBES

Let the given cubic be

$$a_0x^3+3a_1x_2+3a_2x+a_3=0$$

Let $\quad a_0x^3+3a_1x^2+3a_2x+a_3\equiv A(x-p)^3+B(x-q)^3.$ ...(1)

Equating the coefficients of like powers of x on both sides of the identity (1), we get

$$A+B=a_0, \qquad \text{...(2)}$$

$$-(AP+Bq)=a_1 \qquad \text{...(3)}$$

$$Ap^2+Bq^2=a_2 \qquad \text{...(4)}$$

$$-(Ap^3 + Bq^3) = a_3. \quad ..(5)$$

Now, multiplying (2), (3) and (4) by p and adding respectively to (3), (4) and (5), we have

$$B \,.\, (p - q) = a_0p + a_1 \quad ...(6)$$

$$-Bq\,(p - q) = a_1p + a_2 \quad ...(7)$$

and $$Bq^2\,(p - q) = a_2p + a_3 \quad ...(8)$$

From (6), (7) and (8), we have $(a_0p + a_1)(a_2p + a_3) = (a_1p + a_2)^2$. ...(9)

Similarly, multiplying (2), (3) and (4) by q and adding respectively to (3), (4) and (5) and proceeding as above, we have $(a_0q + a_1)(a_2q + a_3) = (a_1q + a_2)^2$. ...(10)

Equations (9) and (10) imply that p and q are the roots of

$$(a_0x + a_1)(a_2x + a_3) = (a_1x + a_2)^2$$

or $$(a_0a_2 - a_1^2)\,x^2 + (a_0a_3 - a_1a_2)\,x + (a_1a_3 - a_2^2) = 0 \quad ...(12)$$

or $$\begin{vmatrix} 1 & -x & x^2 \\ a_0 & a_1 & a_2 \\ a_1 & a_2 & a_3 \end{vmatrix} = 0. \quad ...(13)$$

Having found p and q from above, we can find A and B from (2) and (3), provided $p \neq q$, and the given cubic can then be put as

$$A(x - p)^3 + B(x - q)^3 = 0 \text{ or } A(x - p)^3 = -B\,(x - q)^3$$

or $$\left(\frac{x-p}{x-q}\right)^3 = \left(\frac{-B}{A}\right) \quad ...(14)$$

Thus, if p and q $(p \neq q)$ be the roots of $(a_0x + a_1)(a_2x + a_3) = (a_1x + a_2)^2$, then the cubic equation $a_0x^3 + 3a_1x^2 + 3a_2x + a_3 = 0$ can be reduced to the form $A\,(x - p)^3 + B(x - q)^3 = 0$.

Now, taking cube-roots of both sides of (14), we can find one of the values of x and then depress the equation by one dimension and the remaining two roots can be found from the resulting quadratic.

In case, the values of p and q are imaginary and hence of A, B, then we shall not be able to find the cube-roots of complex quantities. Hence to solve the cubic by the above method, the roots of the quadratic (13) should be real and distinct, the condition for which is $B^2 - 4AC > 0$

i.e., $$(a_0a_3 - a_1a_2)^2 - 4(a_1a_3 - a_2^2)(a_0a_2 - a_1^2) > 0.$$

Hessian: Making the given cubic homogeneous by introducing y, we have

$$f(x, y) \equiv a_0x^3 + 3a_1x^2y + 3a_2xy^2 + a_3y^3 = 0.$$

Then $\dfrac{\partial f}{\partial x} = 3(a_0x^2 + 2a_1xy + a_2y^2);$

$$\frac{\partial^2 f}{\partial x^2} = 6(a_0x + a_1y); \quad \frac{\partial^2 f}{\partial x\,\partial y} = 6(a_1x + a_2y)$$

and $\dfrac{\partial^2 f}{\partial y^2} = 6(a_2x + a_3y).$

Putting y = 1, we see that the quadratic equation (12) whose roots are p and q is given by

$$\frac{\partial^2 f}{\partial x^2} \cdot \frac{\partial^2 f}{\partial y^2} = \left(\frac{\partial^2 f}{\partial x\,\partial y}\right)^2.$$

This equation *i.e.,* the quadratic equation (12) is called the *Hessian* of the cubic. The roots of the Hessian should be real and distinct if the cubic is to be solved by expressing it as the sum or difference of two cubes.

SOLVED EXAMPLES

Example 1:

Show that the roots of the cubic $8a^3x^3 - 6ax + 2\sin 3A = 0$ are

$$\frac{1}{a}\sin A, \frac{1}{a}\sin\left(\frac{\pi}{3} - A\right), -\frac{1}{a}\sin\left(\frac{\pi}{3} + A\right).$$

Solution:

The given cubic can be written as

$$x^3 - \frac{3}{4a^2}x + \frac{1}{4a^3}\sin 3A = 0. \qquad ...(1)$$

Let $x = u^{1/3} + v^{1/3}$ be a root of the given cubic.

Cubing both sides, we get $x^3 = u + v + 3u^{1/3}v^{1/3}(u^{1/3} + v^{1/3})$

or $$x^3 - 3u^{1/3}v^{1/3}x - (u + v) = 0. \qquad ...(2)$$

Comparing (1) and (2), we get $u^{1/3}v^{1/3} = \dfrac{1}{4a^2}$

or $$uv = \frac{1}{64a^6} \text{ and } u + v = -\frac{\sin 3A}{4a^3}.$$

$\therefore$ u and v are the roots of the quadratic $t^2 + \frac{\sin 3A}{4a^3} t + \frac{1}{64a^6} = 0$.

$$t = \frac{-\frac{\sin 3A}{4a^3} \pm \sqrt{\left(\frac{\sin^2 3A}{16a^6} - 4.1.\frac{1}{64a^6}\right)}}{2}$$

$$= \frac{1}{8a^3}\left[-\sin 3A \pm i \cos 3A\right] = \frac{1}{8a^3}\left[\cos\left(\frac{\pi}{2} + 3A\right) \pm i \sin\left(\frac{\pi}{2} + 3A\right)\right].$$

Let $u = \frac{1}{8a^3}\left[\cos\left(\frac{\pi}{2} + 3A\right) + i \sin\left(\frac{\pi}{2} + 3A\right)\right]$

and $v = \frac{1}{8a^3}\left[\cos\left(\frac{\pi}{2} + 3A\right) - i \sin\left(\frac{\pi}{2} + 3A\right)\right]$

Then $x = u^{1/3} + v^{1/3} = \frac{1}{2a}\left[\left\{\cos\left(\frac{\pi}{2} + 3A\right) + i \sin\left(\frac{\pi}{2} + 3A\right)\right\}\right.$

$$\left. + \left\{\cos\left(\frac{\pi}{2} + 3A\right) - i \sin\left(\frac{\pi}{2} + 3A\right)\right\}^{1/3}\right]$$

$$= \frac{1}{2A}\left[2 \cos\frac{1}{3}\left(2n\pi + \frac{\pi}{2} + 3A\right)\right], \text{ where } n = 0, 1, 2.$$

$\therefore$ The roots of the given cubic are

$$\frac{1}{a}\cos\frac{1}{3}\left(\frac{\pi}{2} + 3A\right), \frac{1}{a}\cos\frac{1}{3}\left(2\pi + \frac{\pi}{2} + 3A\right)$$

and $\frac{1}{a}\cos\frac{1}{3}\left(4\pi + \frac{\pi}{2} + 3A\right)$

i.e., $\frac{1}{a}\cos\left(\frac{\pi}{6} + A\right), \frac{1}{a}\cos\left(\frac{5\pi}{6} + A\right)$ and $\frac{1}{a}\cos\left(\frac{3\pi}{2} + A\right)$

i.e., $\frac{1}{a}\cos\left\{\frac{\pi}{2} - \left(\frac{\pi}{3} - A\right)\right\}, \frac{1}{a}\cos\left\{\frac{\pi}{2} + \left(\frac{\pi}{3} + A\right)\right\}$

and $\frac{1}{a}\cos\left\{\frac{\pi}{2} + (\pi + A)\right\}$

i.e., $\frac{1}{a}\sin\left(\frac{\pi}{3} - A\right), -\frac{1}{a}\sin\left(\frac{\pi}{3} + A\right)$ and $\frac{1}{a}\sin A$.

Hence, the roots of the given cubic are

$$\frac{1}{a}\sin A,\ \frac{1}{a}\sin\left(\frac{\pi}{3}-A\right) \text{ and } -\frac{1}{a}\sin\left(\frac{\pi}{3}+A\right).$$

Example 2:

Solve by Cardan's method $27x^3 + 54x^2 + 198x - 73 = 0$.

Solution:

Let us first remove the second term of the given equation by diminishing its roots by h, where $h = -\frac{54}{3\times 27} = -\frac{2}{3}$.

The procedure is as given below.

– 2/3	27	54	198	– 73
		– 18	– 24	– 116
		36	174	– 189
		– 18	– 12	
		18	162	
		– 18		
		0		

∴ The transformed equation is

$$27z^3 + 162z - 189 = 0 \text{ or } z^3 + 6z - 7 = 0. \qquad \text{...(1)}$$

Let $z = u^{1/3} + v^{1/3}$ be a solution of the cubic (1).

Cubing both sides, we get

$$z^3 = u + v + 3u^{1/3}v^{1/3}(u^{1/3} + v^{1/3})$$

or $$z^3 - 3u^{1/3}v^{1/3}z - (u + v) = 0. \qquad \text{...(2)}$$

Comparing (1) and (2), we get $u + v = 7$

and $u^{1/3}v^{1/3} = -2$ or $uv = -8$.

∴ u and v are the roots of the quadratic $t^2 - 7t - 8 = 0$;

giving $t = 8, -1$.

Let $u = 8$ and $v = -1$.

Then $u^{1/3} = 2$ and $v^{1/3} = -1$.

Hence, the roots of the cubic (1) are

$$u^{1/3} + v^{1/3} = 2 - 1 = 1,$$

$$\omega u^{1/3} + \omega^2 v^{1/3} = 2\left(\frac{-1+\sqrt{3}i}{2}\right) - 1\left(\frac{-1-\sqrt{3}i}{2}\right) = -\frac{1}{2} + \frac{3\sqrt{3}}{2}i$$

and $$\omega^2 u^{1/3} + \omega v^{1/3} = 2\left(\frac{-1-\sqrt{3}i}{2}\right) - 1\left(\frac{-1+\sqrt{3}i}{2}\right) = -\frac{1}{2} - \frac{3\sqrt{3}}{2}i.$$

But $z = x + \frac{2}{3}$ or $x = z - \frac{2}{3}$.

Hence, the roots of the given cubic are

$$1 - \frac{2}{3};\ \left(-\frac{1}{2} + \frac{3\sqrt{3}}{2}i\right) - \frac{2}{3};\ \left(-\frac{1}{2} - \frac{3\sqrt{3}}{2}i\right) - \frac{2}{3}$$

i.e., $$\frac{1}{3};\ -\frac{7}{6} + \frac{3\sqrt{3}}{2}i,\ -\frac{7}{6} - \frac{3\sqrt{3}}{2}i.$$ **Ans.**

Example 3:

Solve by Cardan's method $9x^3 + 6x^2 - 1 = 0.$ ...(1)

Solution:

Since the term involving x is missing in the given equation, therefore putting $x = 1/z$, we transform the given equation to the equation whose roots are the reciprocals of the roots of given equation (1).

The transformed equation is,

$$9\left(\frac{1}{z}\right)^3 + 6\left(\frac{1}{z}\right)^2 - 1 = 0$$

or $$z^3 - 6z - 9 = 0. \qquad ...(2)$$

Let $z = u^{1/3} + v^{1/3}$ be a solution of the cubic (2).

Cubing both sides, we get

$$z^3 = u + v + 3u^{1/3}v^{1/3}(u^{1/3} + v^{1/3})$$

or $$z^3 - 3u^{1/3}v^{1/3}z - (u + v) = 0. \qquad ...(3)$$

Comparing (2) and (3), we get

$$u^{1/3}v^{1/3} = 2 \text{ or } uv = 2^3 = 8$$

and $$u + v = 9.$$

$\therefore$ u and v are the roots of the quadratic

$$t^2 - 9t + 8 = 0 \text{ or } (t - 1)(t - 8) = 0. \quad \therefore t = 1, 8.$$

Let $u = 8$ and $v = 1$.

Then, $u^{1/3} = 2$ and $v^{1/3} = 1$.

Hence, the roots of the cubic (2) are

$$u^{1/3} + v^{1/3} = 2 + 1 = 3,$$

$$\omega u^{1/3} + \omega^2 v^{1/3} = 2 \cdot \frac{1}{2}\left(-1+\sqrt{3i}\right)+\frac{1}{2}\left(-1-\sqrt{3i}\right) = \frac{1}{2}\left(-3+\sqrt{3i}\right)$$

and $\omega^2 u^{1/3} + \omega v^{1/3} = 2\cdot\frac{1}{2}\left(-1-\sqrt{3i}\right)+\frac{1}{2}\left(-1+\sqrt{3i}\right) = \frac{1}{2}\left(-3-\sqrt{3i}\right).$

$\therefore$ the roots of the given cubic (1) are

$$\frac{1}{3}, \frac{2}{-3+\sqrt{3i}} \text{ and } \frac{2}{-3-\sqrt{3i}}$$

i.e., $\frac{1}{3}, -\frac{1}{6}\left(3+\sqrt{3i}\right) \text{ and } -\frac{1}{6}\left(3-\sqrt{3i}\right).$ **Ans.**

Example 4:

Prove that the roots of the equation $x^3 - 3x + 1 = 0$ *are*

$$2\cos\frac{2\pi}{9}, 2\cos\frac{8\pi}{9}, 2\cos\frac{14\pi}{9}.$$

Solution:

The given equation is $x^3 - 3x + 1 = 0$ ...(1)

in which the term involving x^2 is missing.

Let $x = u^{1/3} + v^{1/3}$ be a solution of (1).

Cubing both sides, we get $x^3 = u + v + 3u^{1/3}v^{1/3}(u^{1/3} + v^{1/3})$ ...(2)

or $x^3 - 3u^{1/3}v^{1/3}x - (u + v) = 0.$

Comparing (1) and (2), we get

$u^{1/3}v^{1/3} = 1$ or $uv = 1$ and $u + v = -1.$

$\therefore$ u and v are the roots of the quadratic $t^2 + t + 1 = 0.$

$$\therefore\ t = \frac{-1 \pm i\sqrt{3}}{2}.$$

Let $u = -\frac{1}{2} + i\frac{\sqrt{3}}{2}$ and $v = -\frac{1}{2} - i\frac{\sqrt{3}}{2}.$

$$\therefore\ x = \left(-\frac{1}{2}+i\frac{\sqrt{3}}{2}\right)^{1/3} + \left(-\frac{1}{2}-i\frac{\sqrt{3}}{2}\right)^{1/3}.$$

Put $-\frac{1}{2} = r\cos\theta$ and $\frac{\sqrt{3}}{2} = r\sin\theta.$

Then $r^2 = 1$ or $r = 1$ and $\tan\theta = -\sqrt{3}$ or $\theta = \frac{2\pi}{3}$.

$\therefore$ $x = r^{1/3}[(\cos\theta + i\sin\theta)^{1/3} + (\cos\theta - \sin\theta)^{1/3}]$

$$= \left[\left\{\cos\left(\frac{2n\pi+\theta}{3}\right) + i\sin\left(\frac{2n\pi+\theta}{3}\right)\right\} + \left\{\cos\left(\frac{2n\pi+\theta}{3}\right) - i\sin\left(\frac{2n\pi+\theta}{3}\right)\right\}\right]$$

$$= 2\cos\left(\frac{2n\pi}{3} + \frac{2\pi}{9}\right), \text{ where } n = 0, 1, 2.$$

Hence the roots are $2\cos\frac{2\pi}{9}, 2\cos\frac{8\pi}{9}$ and $2\cos\frac{14\pi}{9}$.

Example 5:

Solve the equation $x^3 - 6x - 9 = 0$, by Cardan's Method.

Solution:

The given equation is $x^3 - 6x - 9 = 0$, ...(1)

$\therefore$ Let $x = u^{1/3} + v^{1/3}$, be a solution of the given cubic.

Cubic both sides, we get $x^3 = u + v + 3u^{1/3}v^{1/3}(u^{1/3} + v^{1/3})$

or $x^3 - 3u^{1/3}v^{1/3}x - (u + v) = 0$. ...(2)

Comparing (1) and (2), we get

$u^{1/3}v^{1/3} = 2$ or $uv = 8$ and $u + v = 9$.

$\therefore$ u and v are the roots of the quadratic

u and v are the roots of the quadratic

$t^2 - (u + v)t + uv = 0$

or $t^2 - 9t + 8 = 0$ or $(t - 1)(t - 8) = 0$, giving $t = 1, 8$.

Let $u = 1$ and $v = 8$.

$\therefore$ $u^{1/3} = 1$ and $v^{1/3} = 2$.

Hence the roots of the given equation are $u^{1/3} + v^{1/3} = 1 + 2 = 3$,

$$\omega u^{1/3} + \omega^2 v^{1/3} = 1.\left(-\frac{1}{2} + i\frac{\sqrt{3}}{2}\right) + 2.\left(-\frac{1}{2} - i\frac{\sqrt{3}}{2}\right) = -\frac{3}{2} - \frac{i\sqrt{3}}{2}$$

and $\omega^2 u^{1/3} + \omega v^{1/3} = 1.\left(-\frac{1}{2} - i\frac{\sqrt{3}}{2}\right) - 2.\left(-\frac{1}{2} + i\frac{\sqrt{3}}{2}\right) = -\frac{3}{2} + \frac{i\sqrt{3}}{2}$.

Example 6:

Solve the equation $x^3 - 15x - 126 = 0$, by Cardan's method.

(Meerut, 2003)

Solution:

The given equation is $x^3 - 15x - 126 = 0$, ...(1)

in which the term involving x^2 is missing.

$\therefore$ Let $x = u^{1/3} + v^{1/3}$, be a solution of the given cubic.

Cubing both sides, we get

$$x^3 = u + v + 3u^{1/3}v^{1/3}(u^{1/3} + v^{1/3})$$

or $$x^3 - 3u^{1/3}v^{1/3}x - (u + v) = 0. \qquad ...(2)$$

Comparing (1) and (2), we get

$$u^{1/3}v^{1/3} = 5 \text{ or } uv = 5^3 = 125$$

and $$u + v = 126.$$

$\therefore$ u and v are the roots of the quadratic $t^2 - (u + v)\, t + uv = 0$

or $t^2 - 126t + 125 = 0$ or $(t - 1)(t - 125) = 0$, giving $t = 1, 125$.

Let $u = 1$ and $v = 125$.

$\therefore$ $u^{1/3} = 1$ and $v^{1/3} = 5$.

Hence the roots of the given equation are $u^{1/3} + v^{1/3} = 1 + 5 = 6$,

$$\omega u^{1/3} + \omega^2 v^{1/3} = \left(-\frac{1}{2} + i\frac{\sqrt{3}}{2}\right).1 + \left(-\frac{1}{2} - i\frac{\sqrt{3}}{2}\right).5 = -3 - 2\sqrt{3i}$$

and $$\omega^2 u^{1/3} + \omega v^{1/3} = \left(-\frac{1}{2} - i\frac{\sqrt{3}}{2}\right).1 + \left(-\frac{1}{2} + i\frac{\sqrt{3}}{2}\right).5 = -3 + 2\sqrt{3i}.$$

Example 7:

The given equation is $x^3 + 6x^2 - 12x + 32 = 0$. *...(1)*

Solution:

The given equation is $x^3 + 6x^2 - 12x + 32 = 0$. ...(1)

First we remove the second term of the eqûation (1) by diminishing its roots by

$$h = -\frac{a_1}{na_0} = -\frac{6}{3.1} = -2.$$

The produce is as given below.

– 2	1	6	– 12	32
		– 2	– 8	40
		4	– 20	72
		– 2	– 4	
		2	– 24	
		– 2		
		0		

∴ The transformed equation is $z^3 - 24z + 72 = 0$, ...(1)

where $z = x - (-2) = x + 2$. ...(2)

Let $z = u^{1/3} + v^{1/3}$ be a solution of the cubic (2).

Cubing both sides, we get $z^3 = u + v + 3u^{1/3}v^{1/3}(u^{1/3} + v^{1/3})$

or $\quad z^3 \; 3u^{1/3}v^{1/3}z - (u + v) = 0.$

Comparing (2) and (3), we get $u^{1/3}v^{1/3} = 8$ or $uv = 8^3 = 512$ a nd $u + v = -72$.

∴ u and v are the roots of the quadratic $t^2 - (u + v)\,t + uv = 0$

or $t^2 + 72t + 512 = 0$ or $(t + 8)(t + 64) = 0$, giving $t = -8, -64$.

Let $\quad u = -8$ and $v = -64$.

∴ $\quad u^{1/3} = -2$ and $v^{1/3} = -4$.

∴ Roots of the cubic (2), are $u^{1/3} + v^{1/3} = -2 - 4 = -6$.

$$wu^{1/3} + w^2v^{1/3} = \left(-\frac{1}{2}+i\frac{\sqrt{3}}{2}\right).(-2)+\left(-\frac{1}{2}-i\frac{\sqrt{3}}{2}\right).(-4) = 3+i\sqrt{3}$$

$$\text{and } w^2u^{1/3} + wv^{1/3} = \left(-\frac{1}{2}-i\frac{\sqrt{3}}{2}\right).(-2)+\left(-\frac{1}{2}+i\frac{\sqrt{3}}{2}\right).(-4)$$

$$= 3 - i\sqrt{3}.$$

But $x = z - 2$.

Hence, the roots of the given cubic are $-6 - 2$, $3 + i\sqrt{3} - 2$ and $3 - i\sqrt{3} - 2$

or -8, $1 + i\sqrt{3}$ and $1 - i\sqrt{3}$.

Example 8:

Solve by the Cardan's method $x^3 - 15x^2 - 33x + 847 = 0.$

(Meerut, 1991)

Solution:

The given equation is $x^3 - 15x^2 - 33x + 847 = 0$. ...(1)

Let us first remove the second term of the given equation by diminishing its roots by

$$h = -\frac{a_1}{na_0} = -\frac{-15}{3.1} = 5.$$

The procedure is as given below.

5	1	– 15	– 33	847
		5	– 50	– 415
		– 10	– 83	432
		5	– 25	
		– 5	– 108	
		5		
		0		

$\therefore$ The transformed equation is $z^3 - 108z + 432 = 0$

where $z = x - 5$.

Let $z = u^{1/3} + v^{1/3}$ be a solution of the cubic (2).

Cubing both sides, we get $z^3 = u + v + 3u^{1/3}v^{1/3}(u^{1/3} + v^{1/3})$

or $\quad z^3 - 3u^{1/3}v^{1/3}z - (u + v) = 0$.

Comparing (2) and (3), we get $u^{1/3}v^{1/3} = 36$ or $uv = 36^3$

and $\quad u + v = -432$.

$\therefore$ u and v are the roots of the quadratic $t^2 + 432t + 36^3 = 0$.

Solving this, we get

$$t = \frac{-432 \pm \sqrt{[432]^2 - 4(36)^3}}{2} = \frac{-432 \pm 0}{2} = -216.$$

$\therefore \quad u = v = -216$.

$\therefore \quad u^{1/3} = -6$ and $v^{1/3} = -6$.

$\therefore$ the roots of the cubic in z are

$$u^{1/3} + v^{1/3},\ \omega u^{1/3} + \omega^2 v^{1/3} \text{ and } \omega^2 u^{1/3} + \omega v^{1/3}$$

i.e., $\quad -12,\ -6(\omega + \omega^2)$ and $-6(\omega^2 + \omega)$

i.e., $\quad -12$, 6 and 6, since $\omega + \omega^2 = -1$.

Hence, the roots of the given cubic are

$-12+5, 6+5, 6+5$ *i.e.*, $-7, 11, 11$. **Ans.**

Example 9:

Solve the cubic $x^3+6x^2+9x+4=0$ *by Cardan's method.*

(Meerut, 1989)

Solution:

The given equation is $x^3+6x^2+9x+4=0$. ...(1)

First we remove the second term of the equation (1) by diminishing its roots by

$$h=-\frac{a_1}{na_0}=-\frac{6}{3.1}=-2.$$

The procedure is as given below.

−2	1	6	9	4
		−2	−8	−2
		4	1	2
		−2	−4	
		2		−3
		−2		
		0		

∴ The transformed equation is $z^3-3z+2=0$, ...(2)

where $z=x-(-2)=x+2$

Let $z=u^{1/3}+v^{1/3}$ be a solution of the cubic (2).

Cubing with sides, we get $z^3=u+v+3u^{1/3}v^{1/3}(u^{1/3}+v^{1/3})$

or $z^3-3u^{1/3}v^{1/3}z-(u+v)=0$. ...(3)

Comparing (2) and (3), we get $u^{1/3}v^{1/3}=1$ or $uv=1$ and $u+v=-2$.

∴ u and v are the roots of the quadratic $t^2-(u+v)t+uv=0$

or $t^2+2t+1=0$

or $(t+1)^2=0$, giving $t=-1, -1$.

∴ $u=v=-1$

or $u^{1/3}=v^{1/3}=-1$.

∴ Roots of the cubic (2), are $u^{1/3}+v^{1/3}=-1-1=-2$

$$\omega u^{1/3} + \omega^2 v^{1/3} = (\omega + \omega^2)(-1) = (-1)(-1) = 1$$

$$[\because 1 + \omega + \omega^2 = 0]$$

and $\omega^2 u^{1/3} + \omega v^{1/3} = (\omega + \omega^2)(-1) = (-1)(-1) = 1.$

But $x = z - 2$. Hence the roots of the given cubic (1) are

$-2 - 2,\ 1 - 2$ and $1 - 2$

i.e., -4, -1 and -1.

Example 10:

If the cubic equation $a_0x^3 + 3a_1x^2 + 3a_2x + a_3 = 0$ *has two roots equal to* α*, prove that*

$$-\alpha = \frac{H_2}{H_1} = \frac{H_1}{H},$$

where $H = a_0a_2 - a_1^2,\ 2H_1 = a_0a_3 - a_1a_2,$
$H_2 = a_1a_3 - a_2^2.$

Solution:

We know that if a cubic has two roots equal, then its Hessian has also two equal roots in common with the cubic. The Hessian of given cubic is

$$(a_0a_2 - a_1^2)x^2 + (a_0a_3 - a_1a_2)x + (a_1a_3 - a_2^2) = 0$$

i.e., $\quad Hx^2 + 2H_1x + H_2 = 0 \qquad ...(1)$

Since Hessian has equal roots, therefore we have

$$4H_1^2 - 4HH_2 = 0$$

or $\quad H_1^2 = HH_2. \qquad ...(2)$

But each of the equal roots is common with that of the cubic and hence it is also α.

$\therefore\ 2\alpha =$ sum of the roots] of (1) $= -\dfrac{2H_1}{H}$

or $\quad -\alpha = \dfrac{H_1}{H}.$

Also from (2), we have $-\alpha = \dfrac{H_1}{H} = \dfrac{H_2}{H_1}.$

Example 11:

Solve the equation $x^3 - 3x^2 + 33x - 1 = 0$ *by expressing it as the sum or difference of two cubes.*

Solution:

Let $x^3 - 3x^2 + 33x - 1 \equiv A(x-p)^3 - (x-q)^3 = 0$ ...(1)

Then p and q are the roots of $\begin{vmatrix} 1 & -x & x^2 \\ a_0 & a_1 & a_2 \\ a_1 & a_2 & a_3 \end{vmatrix} = 0$,

where $a_0 = 1$, $a_1 = 1$, $a_2 = 11$, $a_3 = -1$.

Thus p and q are the roots of the quadratic $\begin{vmatrix} 1 & -x & x^2 \\ 1 & -1 & 11 \\ -1 & 11 & -1 \end{vmatrix} = 0$.

Expanding the determinant, we get $10(x^2 + x - 12) = 0$

or $x^2 + x - 12 = 0$.

$\therefore$ $x = 3, -4$ and so $p = 3$ and $q = -4$.

Comparing the coefficients of x^3 and x^2 in (1), we have $A - B = 1$ and $Ap - Bq = 1$ *i.e.,* $3A + 4B = 1$.

Solving these equations, we get $A = \frac{5}{7}, B = -\frac{2}{7}$.

Hence, from (1), the given cubic may be written as

$$\frac{5}{7}(x-3)^3 + \frac{2}{7}(x+4)^3 = 0$$

or $5(x-3)^3 = -2(x+4)^3$ or $\left(\frac{x-3}{x+4}\right)^3 = -\frac{2}{5}$.

$\therefore$ $\frac{x-3}{x+4} = -\left(\frac{2}{5}\right)^{1/3} k$, where $k = 1, \omega, \omega^2$ and $k^3 = 1$

or $\frac{x-3}{x+4} = -\frac{a}{b}k$, where $a = 2^{1/3}$, $b = 5^{1/3}$

or $x(b + ak) = 3b - 4ak$ or $x = \frac{3b - 4ak}{b + ak} \cdot \frac{b^2 - abk + a^2k^2}{b^2 - abk + a^2k^2}$

[Multiplying the Nr. and the Dr. by $b^2 - abk + a^2k^2$ so that the Dr. becomes $b^3 + a^3k^3$]

$$= \frac{(3b^3 - 4a^3k^3) - k(3ab^2 + 4ab^2) + k^2(4a^2b + 3ba^2)}{b^3 + a^3k^3}$$

$$= \frac{(15-8) - 7k2^{1/3}.5^{2/3} + 7k^2.2^{2/3}.5^{1/3}}{5+2} \qquad [\because k^3 = 1]$$

Thus, the required roots are $\frac{7 - 7k.2^{1/3}.5^{2/3} + 7k^2.2^{2/3}.5^{1/3}}{7}$,

where k = 1, ω, ω^2 respectively.

EXERCISES

1. $x^3 + 3ax^2 + 3(a^2 - bc)x + a^3 + b^3 + c^3 - 3abc = 0$. **(Meerut, 1988)**
2. $x^3 - 3a^2x - 2a^3\cos 3A = 0$. **(Meerut, 1986)**
3. $64x^3 - 144x^2 + 108x - 27 = 0$.
4. $x^3 + 3x^2 - 27x + 104 = 0$. **(Meerut, 1993)**
5. $x^3 - 12x^2 - 6x - 10 = 0$.
6. $28x^3 - 9x^2 + 1 = 0$. **(Meerut, 1988, 2003)**
7. $2x^3 + 3x^2 + 3x + 1 = 0$. **(Meerut, 1996)**
8. $x^3 - 18x - 35 = 0$. **(Meerut, 1992)**
9. $x^3 - 15x^2 - 357x + 5491 = 0$.
10. $x^3 - 21x - 344 = 0$.
11. $152x^3 - 60x^2 - 606x - 485 = 0$.
12. $2x^3 + 3x^2 - 21x + 19 = 0$.
13. $9x^3 - 30x^2 + 36x - 16 = 0$.
14. $x^3 + 3x^2 - 27x + 104 = 0$.
15. Show that if p and q ($p \neq q$) be the roots of $(a_0x + a_1)(a_2x + a_3) = (a_1x + a_2)^2$, then the cubic equation $a_0x^3 + 3a_1x^2 + 3a_2x + a_3 = 0$ can be reduced to the form

    $A(x - p)^3 + B(x - q)^3 = 0$.

    Solve the following equations by expressing them as the sum or difference of two cubes.

5

SOLUTION OF BIQUADRATIC EQUATIONS

5.1 EULER'S SOLUTION OF THE BIQUADRATIC

Let the biquadratic $a_0x^4 + 4a_1x^3 + 6a_2x^2 + 4a_3x + a_4 = 0$ ...(1)

be put in the form $z^4 + 6Hz^2 + 4Gz + (a_0^2I - 3H^2) = 0,$ ...(2)

where $z = a_0x + a_1.$

Let $z = \sqrt{p} + \sqrt{q} + \sqrt{r}.$ ...(3)

Squaring, we get $[z^2 - (p + q + r)] = 2\left(\sqrt{p}\sqrt{q} + \sqrt{q}\sqrt{r} + \sqrt{r}\sqrt{p}\right).$

Again squaring both sides, we get

$$z^4 - 2z^2(p + q + r) + (p + q + r)^2 = \left[pq + qr + rp + 2\sqrt{p}\sqrt{q}\sqrt{r}\left(\sqrt{p} + \sqrt{q} + \sqrt{r}\right)\right]$$

or $z^4 - (2\ S\ p)\ z^2 - 8z\ \sqrt{p}\sqrt{q}\sqrt{r} + (\Sigma p)^2 - 4\Sigma pq = 0.$ ...(4)

Comparing the coefficients of like powers of z in (2) and (4), we get

$$\Sigma p = -3H, \sqrt{p}\sqrt{q}\sqrt{r} = -\frac{G}{2}$$

or $$\Sigma pq = 3H^2 - \frac{a_0^2 I}{4}.$$

Hence, the equation whose roots are p, q, r is $t^3 - (\Sigma p)\ t^2 + (\Sigma pq)\ t - pqr = 0$

or $$t^3 + 3Ht^2 + \left(3H^2 - \frac{a_0^2 I}{4}\right)t - \frac{G^2}{4} = 0. \qquad ...(5)$$

This equation is known as *Euler's cubic.*

We know that $G^2 + 4H^3 = a_0(HI - a_0J)$.

$$\therefore \quad \frac{-G^2}{4} = H^3 - \frac{a_0^2HI}{4} + \frac{a_0^3J}{4}. \qquad ...(6)$$

Hence from (5), we have

$$t^3 + 3Ht^2 + 3H^2t - \frac{a_0^2I}{4}t + H^3 - \frac{a_0^2HI}{4} + \frac{a_0^3J}{4} = 0$$

or
$$(t+H)^3 - \frac{a_0^2I}{4}(t+H) + \frac{a_0^3J}{4} = 0.$$

Putting $t + H = a_0^2\theta$, we have

$$a_0^6\theta^3 - \frac{a_0^4I}{4}\theta + \frac{a_0^3J}{4} = 0$$

or
$$4a_0^3\theta^3 - Ia_0\theta + J = 0. \qquad ...(6)$$

This equation is called *reducing cubic* of the biquadratic equation.

Remark:

Since in Euler's cubic, G occurs in even powers, so if we had the biquadratic as

$$z^4 + 6Hz^2 - 4Gz + \left(a_0^2I - 3H^2\right) = 0,$$

then its Euler's cubic would have been the same as that of the given biquadratic and therefore the two will have the same reducing cubic.

5.2 RELATIONS BETWEEN THE ROOTS OF THE BIQUADRATIC AND EULER'S CUBIC

We have taken $z = \sqrt{p} + \sqrt{q} + \sqrt{r}$. ...(1)

When each of the three radicals have their double sign, we have eight values of z by the above assumption, but there is a condition that

$$\sqrt{p}\sqrt{q}\sqrt{r} = -\frac{G}{2}$$

or
$$\sqrt{r} = \frac{-G}{2\sqrt{p}\sqrt{q}} \qquad ...(2)$$

$$\therefore \quad z = \sqrt{p} + \sqrt{q} - \frac{G}{2\sqrt{p}\sqrt{q}}.$$

From this we get four values of z, since $\sqrt{p}, \sqrt{q}$ have double signs.

The signs to be affixed before $\sqrt{p}, \sqrt{q}, \sqrt{r}$ should be such as to satisfy the condition (2). If G is negative, *i.e.,* $-\frac{G}{2}$ is +ive, we should fix such signs as to make $\sqrt{p}\sqrt{q}\sqrt{r}$ also +ive. Now, let z_1, z_2, z_3, z_4 be the values of z and $\alpha, \beta, \gamma, \delta$ be the roots of the given biquadratic, then we have

$$\left.\begin{aligned} z_1 &= a_0\alpha = \sqrt{p}-\sqrt{q}-\sqrt{r}, && \sqrt{p}+\text{ive} \\ z_2 &= a_0\beta = -\sqrt{p}+\sqrt{q}-\sqrt{r}, && \sqrt{q}+\text{ive} \\ z_3 &= a_0\gamma + a_1 = -\sqrt{p}-\sqrt{q}+\sqrt{r}, && \sqrt{r}+\text{ive} \\ z_4 &= a_0\delta + a_1 = \sqrt{p}+\sqrt{q}+\sqrt{r}, && \text{all}+\text{ive} \end{aligned}\right\} \quad ...(1)$$

$\therefore$ $a_0(\alpha + \beta) + 2a_1 = -2\sqrt{r}$, $a_0(\gamma + \delta) + 2a_1$

$$= 2\sqrt{r},\ a_0(\alpha + \beta - \gamma - \delta) = -4\sqrt{r}$$

or $\qquad r = \frac{a_0^2}{16}(\alpha + \beta - \gamma - \delta)^2.$

Similarly, we have

$$q = \frac{a_0^2}{16}(\gamma + \alpha - \beta - \delta)^2,$$

$$p = \frac{a_0^2}{16}(\beta + \gamma - \alpha - \delta)^2.$$

In case G is +ive, *i.e.,* $-\frac{G}{2}$ is –ive, then we shall choose all –ive or two +ive and one –ive, *i.e.,* $z_1 = a_0\alpha + a_1 = -\sqrt{p}+\sqrt{q}+\sqrt{r}$, $\sqrt{p}$ is –ive; etc.

5.3 RELATIONS BETWEEN THE ROOTS OF THE BIQUADRATIC AND THE REDUCING CUBIC

The reducing cube is

$$4a_0^3\theta^3 - Ia_0\theta + J = 0$$

From the relations (I) of § 5.2, we have $a_0(\alpha - \beta) = 2\left(\sqrt{p}-\sqrt{q}\right)$ and $-a_0(\gamma - \delta) = 2\left(\sqrt{p}+\sqrt{q}\right)$.

$$\therefore\ -a_0^2(\alpha-\beta)(\gamma-\delta) = 4(p-q) = 4a_0^2(\theta_1-\theta_2) \quad [\because t + H = a_0^2\theta]$$

$$\left.\begin{aligned}4(p-q) &= 4a_0^2(\theta_1-\theta_2) = -a_0^2(\alpha-\beta)(\gamma-\delta)\\ 4(q-r) &= 4a_0^2(\theta_2-\theta_3) = -a_0^2(\beta-\gamma)(\alpha-\delta)\\ 4(r-p) &= 4a_0^2(\theta_3-\theta_1) = -a_0^2(\gamma-\alpha)(\beta-\delta)\end{aligned}\right\} \quad \text{...(I)}$$

Subtracting first relation of (I) from the third, we get

$$4a_0^2[\theta_3 - \theta_1 - \theta_1 + \theta_2] = -a_0^2[(\gamma-\alpha)(\beta-\delta) - (\alpha-\beta)(\gamma-\delta)].$$

Putting $\theta_1 + \theta_2 + \theta_3 = 0$ *i.e.,* $\theta_2 + \theta_3 = -\theta_1$, we get

$$12\theta_1 = (\gamma-\alpha)(\beta-\delta) - (\alpha-\beta)(\gamma-\delta).$$

Similarly, $12\theta_2 = (\alpha-\beta)(\gamma-\delta) - (\beta-\gamma)(\alpha-\delta)$ ...(II)

and $12\theta_3 = (\beta-\gamma)(\alpha-\delta) - (\gamma-\alpha)(\beta-\delta)$

5.4 DESCARTE'S METHOD OF SOLVING A BIQUADRATIC

By this method we resolve the biquadratic into quadratic factors.

Let the given equation be

$$f(x) \equiv a_0x^4 + 4a_1x^3 + 6a_2x^2 + 4a_3x + a_4 = 0.$$

Removing the second term and multiplying the roots by a_0 the above equation is reduced to the form

$$f(z) \equiv z^4 + 6Hz^2 + 4Gz + a_0^2I - 3H^2 = 0,$$

where $z = a_0x + a_1$.

If z_1, z_2, z_3, z_4 are the roots of the equation $f(z) = 0$,

then $z_1 + z_2 + z_3 + z_4 = 0$

or $(z_1 + z_2) = -(z_3 + z_4) = p$ (say).

Further let $z_1z_2 = q$ and $z_3z_4 = q'$.

$\therefore\ f(z) \equiv (z^2 - pz + q)(z^2 + pz + q')$

i.e., $z^4 + 6Hz^2 + 4Gz + a_0^2I - 3H^2 \equiv (z^2 - pz + q)(z^2 + pz + q')$.

Comparing the coefficients of like powers of z, we have

$p + q' - p^2 = 6H,$ ...(1)

$p(q - q') = 4G,$ ...(2)

and $qq' = a_0^2I - 3H^2$ ...(3)

But we have, $(q - q')^2 = (q + q')^2 - 4qq'$.

$$\therefore\ \frac{16G^2}{p^2} = \left(p^2 + 6H\right)^2 - 4\left(a_0^2I - 3H^2\right)$$

or $\quad p^2(p^4 + 12Hp^2 + 36H^2) - 4p^2(a_0{}^2I - 3H^2) - 16G^2 = 0$

or $\quad p^6 + 12Hp^4 + 4p^2(12H^2 - a_0{}^2I) - 16G^2 = 0.$

This being a cubic equation in p^2 can be solved generally by trial method and having found p^2 we can find the values of q and q'. Thus, the quadratic factors of f(z) are known. Hence, we can find the roots of f(z) = 0 and consequently of f(x) = 0.

The above cubic equation in p^2 can be put as

$(p^6 + 12Hp^4 + 48p^2H^2 + 64H^3) - 4a_0{}^2I(p^2 + 4H) - 16G^2 - 64H^3 + 16a_0{}^2IH = 0$

or $(p^2 + 4H)^3 - 4a_0{}^2I(p^2 + 4H) - 16(-a_0{}^3J) = 0.$

$$\left[\because G^2 + 4H^3 = a_0^2 HI = a_0^3 J\right]$$

Putting $p^2 + 4H = 4a_0^2\theta$, we get $4a_0^3\theta^3 - Ia_0\theta + J = 0$.

This equation is called the reducing cubic and is the same as found earlier.

To find the value of p^2 by trial, we try those numbers only which are whole squares, *i.e.*, 1, 4, 9, 16 etc. and then see if they satisfy the cubic.

5.5 FERRARI'S METHOD OF SOLVING THE BIQUADRATIC EQUATIONS

Let the given biquadratic be

$$f(x) \equiv ax^4 + 4bx^3 + 6cx^2 + 4dx + e = 0.$$

Let us suppose that

$$f(x) \equiv \frac{1}{a}\{(ax^2 + 2bx + c + 2aq)^2 - (2Mx + N)^2\}.$$

Comparing the coefficients of like powers of x, we get

$$M^2 = b^2 - ac + a^2\theta. \qquad \text{...(1)}$$

$$MN = bc - ad + 2ab\,\theta. \qquad \text{...(2)}$$

$$N^2 = (c + 2a\theta)^2 - ae. \qquad \text{...(3)}$$

$\therefore (b^2 - ac + a^2\theta)\{(c + 2a\theta)^2 - ae\} = (bc - ad + 2ab\,\theta)^2$

$$[\because M^2N^2 = (MN)^2]$$

or $4a^3\theta^3 - (ae - 4bd + 3c^3)\,a\theta + ace + 2bcd - ad^2 - eb^2 - c^3 = 0$

or $4a^3\theta^3 - Ia\theta + J = 0. \qquad \text{...(4)}$

Equation (4) is the *reducing cubic.* From this equation we can obtain the value of θ. Hence the values of M and N are obtained. Thus, the given biquadratic can be expressed as difference of two squares.

Clearly, the given biquadratic equation reduces to the quadratic equations

$$ax^2 + 2(b - M)x + c + 2a\theta - N = 0 \qquad ...(5)$$

and $$ax^2 + 2(b + M)x + c + 2a\theta + N = 0. \qquad ...(6)$$

Let $\theta_1, \theta_2, \theta_3$ be the three values of θ satisfying (4) and the corresponding values of M be M_1, M_2, M_3 and those of N be N_1, N_2, N_3.

Let β, γ be the roots of (5) and α, δ be those of (6) when M, N, θ have the values of M_1, N_1 and θ_1 respectively. Then

$$\beta+\gamma = -\frac{2}{a}(b-M_1) \text{ and } \beta\gamma = \frac{c+2a\theta_1 - N_1}{a},$$

$$\alpha+\delta = -\frac{2}{a}(b+M_1) \text{ and } \alpha\delta = \frac{c+2a\theta_1 + N_1}{a}. \qquad ...(A)$$

$$\therefore \quad \beta + \gamma - \alpha - \delta = \frac{4}{a}M_1.$$

Similarly, $$\gamma+\alpha-\beta-\delta = \frac{4}{a}M_2 \qquad ...(i)$$

and $$\alpha+\beta-\gamma-\delta = \frac{4}{a}M_3$$

$$\beta\gamma+\alpha\delta = \frac{2c}{a}+4\theta_1$$

$$\gamma\alpha+\beta\delta = \frac{2c}{a}+4\theta_1$$

and $$\gamma\alpha+\beta\delta = \frac{2c}{a}+4\theta_2 \qquad ...(ii)$$

$$\alpha\beta+\gamma\delta = \frac{2c}{a}+4\theta_3$$

From (I), we get

$$-\frac{4}{a}(M_2+M_3-M_1) = (2\alpha-2\delta-\beta-\gamma+\alpha+\delta)$$

$$= (3\alpha - \beta - \gamma - \delta) = (4\alpha - \Sigma\alpha)$$

$$= \left(4\alpha + \frac{4b}{a}\right) = \frac{4}{a}(a\alpha + b).$$

Also from (I) and from (A), we have

$$a\alpha + b = -M_1 + M_2 + M_3$$
$$a\beta + b = M_1 - M_2 + M_3$$
$$a\gamma + b = M_1 + M_2 - M_3$$
$$a\delta + b = -M_1 - M_2 - M_3$$

$$\beta\gamma - \alpha\delta = \frac{-2N_1}{a}$$

and $$\gamma\alpha - \beta\delta = \frac{-2N_2}{a} \qquad \text{...(iii)}$$

$$\alpha\beta - \gamma\delta = \frac{-2N_3}{a}.$$

SOLVED EXAMPLES

Example 1:

Solve the equation $x^4 - 8x^3 - 12x^2 + 60x + 63 = 0$ by Ferrari's method.

Solution:

Given biquadratic can be rewritten as $x^4 - 8x^3 = 12x^2 - 60x - 63$.

Adding $16x^2$ to both sides, we get

$$x^4 - 8x^3 + 16x^2 = 28x^2 - 60x - 63$$

or $$(x^2 - 4x)^2 = 28x^2 - 60x - 63.$$

Then $$(x^2 - 4x + \lambda)^2 = 28x^2 - 60x - 63 + 2\lambda(x^2 - 4x) + \lambda^2$$

or $$(x^2 - 4x + \lambda)^2 = (28 + 2\lambda)x^2 - (60 + 8\lambda)x + (\lambda^2 - 63).$$

Put $$28 + 2\lambda = 1, 4, 9;$$

i.e., $$\lambda = -\frac{27}{4}, -12, -\frac{19}{2}.$$

For $\lambda = -12$, we have $28 + 2\lambda = 4$, $\lambda^2 - 63 = 144 - 63 = 81$

and $60 + 8\lambda = 60 - 96 = -36$.

$\therefore$ The given equation reduces to

$$(x^2 - 4x - 12)^2 = 4x^2 + 36x + 81 = (2x + 9)^2$$

or $$(x^2 - 4x - 12)^2 - (2x + 9)^2 = 0$$

or $$(x^2 - 4x - 12 + 2x + 9)(x^2 - 4x - 12 - 2x - 9) = 0$$

or $$(x^2 - 2x - 3)(x^2 - 6x - 21) = 0.$$

$\therefore \quad x^2 - 2x - 3 = 0$ or $x^2 - 6x - 21 = 0.$

Solving, we get $x = 3 \pm \sqrt{30}, 3, -1.$

Example 2:

Solve the equation $x^4 + 2x^3 - 7x^2 - 8x + 12 = 0$, by Ferrari's Method.

Solution:

The given equation can be rewritten as $x^4 + 2x^3 = 7x^2 + 8x - 12.$

Adding x^2 to both sides to make the L.H.S. a perfect square, we get

$$(x^2 + x)^2 = 8x^2 + 8x - 12.$$

$$\therefore \quad (x^2 + x + \lambda)^2 = 8x - 12 + \lambda^2 + 2\lambda(x^2 + x)$$

$$= x^2(8 + 2\lambda) + x(8 + 2\lambda) + \lambda^2 - 12. \qquad ...(1)$$

If R.H.S. is a perfect square *i.e.*, $B^2 - 4AC = 0$, then we get a cubic in λ from which λ can be found.

The solution of cubic in λ may not be easy and as such we try other method. If R.H.S. is to be a perfect square, then the coefficient of x^2 as well as constant term both should be perfect squares. We put the coefficient of x^2 equal to $1, 4, 9, 16, \frac{1}{4}, \frac{1}{9}, \frac{1}{16}$ etc. and see that the value of l thus obtained makes the constant term also a perfect square and also a +ive quantity.

In this case, let $8 + 2\lambda = 1$ *i.e.*, $\lambda = -\frac{7}{2}$. This value of λ makes $\lambda^2 - 12 = \frac{1}{4}$, which is a perfect square. Clearly, for $\lambda = -7/2$, the R.H.S. of (1) $= \left(x+\frac{1}{2}\right)^2.$

$$\therefore \quad \left(x^2 + x - \frac{7}{2}\right)^2 = \left(x + \frac{1}{2}\right)^2.$$

Thus $\quad f(x) \equiv \left(x^2 + x - \frac{7}{2}\right)^2 - \left(x + \frac{1}{2}\right)^2 = 0$

i.e., $(x^2 + 2x - 3)(x^2 - 4 = 0$ *i.e.*, $x = \pm 2, -3, 1.$

Example 3:

Solve the equation $x^4 - 6x^3 - 9x^2 + 66x - 22 = 0$ by Descarte's Method.

Solution:

To remove the 2nd term, we have to diminish the roots by $\frac{3}{2}$ which will be difficult. So, we first multiply the roots by 2.

Multiplying the roots by 2, the reduced equation is

$$y^4 - 12y^3 - 36y^2 + 528y - 352 = 0, \text{ where } y = 2x.$$

We now diminish the roots by 3.

3	1	– 12	– 36	528	– 352
		3	– 27	– 189	1017
		– 9	– 63	329	665
		3	– 18	– 243	
		– 6	– 81	96	
		3	– 9		
		– 3	– 90		
		3			
		0			

$\therefore$ $f(z) \equiv z^4 - 90z^2 + 96z + 665 = 0,$

where $z = y - 3 = 2x - 3.$

Let $f(z) \equiv (z^2 + pz + q)(z^2 - pz + q').$

Comparing the coefficients, we get $q + q' - p^2 = -90$

and $p(q' - q) = 96$ *i.e.,* $q' - q = \frac{96}{p}.$

Also $qq' = 665.$

Now $(q' - q)^2 = (q + q')^2 - 4qq'$

or $\left(\frac{96}{p}\right)^2 = (p^2 - 90)^2 - 4\,(665).$

Putting $p^2 = t$, we get the t-cubic as

$$t^3 - 180t^2 + 20 \times 272t - 96^2 = 0.$$

Bt trial $t = p^2 = 36$ satisfies this and so $p = 6.$

Putting $p = 6$, we get $q + q' = -54$ and $q' - q = .16.$

$\therefore$ $q = -35$ and $q' = -19$. These values of q and q' also satisfy the relation $qq' = 665.$

$\therefore \quad f(z) \equiv (z^2 + 6z - 35)(z^2 - 6z - 19) = 0.$

Thus $\quad z = \dfrac{-6 \pm \sqrt{(36+140)}}{2}, \dfrac{6 \pm \sqrt{(36+76)}}{2}.$

Hence $\quad 2x - 3 = -3 \pm 2\sqrt{11},\ 3 \pm 2\sqrt{7}$

or $\quad x = \pm\sqrt{11}, 3 \pm \sqrt{7}.$

Example 4:

Solve the equation $x^4 - 3x^2 - 42x - 40 = 0$ Descarte's method.

Solution:

The given equation is $f(x) \equiv x^4 - 3x^2 - 42x - 40 = 0.$

Let $f(x) \equiv (x^2 + pq + q)(x^2 - px + q').$

Comparing the coefficients, we get

$$q + q' - p^2 = -3 \text{ or } q + q' = p^2 - 3;$$

$$p(q' - q) = -42 \text{ or } q' - q = -\frac{42}{p}$$

and $\quad qq' = -40.$

Now $\quad (q' - q)^2 = (q + q')^2 - 4pq'$

or $\quad \dfrac{42^2}{p^2} = \left(p^2 - 3\right)^2 + 160$

or $t^3 - 6t^2 + 169t - 1764 = 0$, where $t = p^2$.

Clearly, $t = 9$ satisfies the above cubic.

Hence $p^2 = 9$ or $p = 3, -3$.

Taking $p = 3$, we get $q + q' = 6$ and $q' - q = -14$.

These given $q' = -4$ and $q = 10$ and these values satisfy the relation

$qq' = -40.$

$\therefore f(x) \equiv (x^2 + 3x + 10)(x^2 - 3x - 4) = 0.$

Hence $x = 4, -1, \dfrac{-2 \pm i\sqrt{31}}{2}.$

Example 5:

Solve the biquadratic equation $x^4 - 8x^3 - 12x^2 + 60x + 63 = 0$ by Descarte's method.

Solution:

The given biquadratic is $x^4 - 8x^3 - 12x^2 + 60x + 63 = 0$ ...(1)

Let us first remove the second term of the given equation (1) by diminishing its roots by h where

$$h = -\frac{a_1}{n a_0} = -\frac{-8}{4\times 1} = 2. \qquad ...(2)$$

The procedure is as follows.

2	1	– 8	– 12	60	63
		2	– 12	– 48	24
		– 6	– 24	12	87
		2	– 8	– 64	
		– 4	– 32	– 52	
		2	– 4		
		– 2	– 36		
		2			
		0			

$\therefore$ The transformed equation is $f(z) \equiv z^4 - 36z^2 - 52z + 87 = 0$, where $z = x - 2$.

Now, let $z^4 - 36z^2 - 52z + 87 \equiv (z^2 + pz + q)(z^2 - pz + q')$ ...(3)

Equating the coefficients of like powers of z on both sides of the identity (3), we get

$$q + q' - p^2 = -36 \Rightarrow q + q' = p^2 - 36$$

$$p(q' - q) = -52 \Rightarrow q' - q = -\frac{52}{p}$$

and $qq' = 87$

Now, $(q' - q)^2 = (q' + q)^2 - 4qq'$.

$$\therefore \quad \frac{52^2}{p^2} = \left(p^2 - 36\right)^2 - 348$$

or $\quad t(t - 36)^2 - 348t - 52 \times 52 = 0$, where $t = p^2$

or $\quad t^3 - 72t^2 + 948t - 2704 = 0.$

Clearly, $t = 4$ satisfies the above cubic in t.

$\therefore t = p^2 = 4.$

Choosing $p = 2$, we get $q + q' = -32$ and $q' - q = -26$

Solving these, we get $q = -3$ and $q' = -29$.

$\therefore$ The equation f(z) = 0 reduces to

$$(z^2 + 2z - 3)(z^2 - 2z - 29) = 0$$

$\Rightarrow$ $\quad z^2 + 2z - 3 = 0$ or $z^2 - 2z - 29 = 0.$

Now $\quad z^2 + 2z - 3 = 0 \Rightarrow (z + 3)(z - 1) = 0$

$\Rightarrow$ $\quad z = -3$ or $z = 1.$

Again $z^2 - 2z - 29 = 0$ gives $z = \dfrac{2 \pm \sqrt{(4+116)}}{2} = 1 \pm \sqrt{30}.$

$\therefore$ $\quad z = -3, 1, 1 \pm \sqrt{30}$

Hence $x = z + 2 = -1, 3, 3 \pm \sqrt{30}.$

EXERCISES

Solve the following biquadratic equations by Ferrari's method *i.e.*, by expressing them as the difference of two squares.

1. Resolve into quadratic factors $x^4 + 3x^3 + x^2 = 2.$
2. $x^4 + 12x - 5 = 0.$
3. $x^4 + 4x^3 + 12x^2 - 8x + 95 = 0.$
4. $x^4 - 2x^3 - 5x^2 + 10x - 3 = 0.$
5. $x^4 - 8x^3 - 12x^2 + 60x + 63 = 0.$
6. $x^4 - 10x^3 + 35x^2 - 50x + 24 = 0.$
7. $2x^4 + 6x^3 - 3x^2 + 2 = 0.$ **(Meerut, 2003)**
8. $x^4 - 10x^3 + 44x^2 - 104x + 96 = 0.$

Solve the following biquadratic equations by Descarte's Method.

9. $x^4 - 3x^2 - 6x - 2 = 0.$
10. $x^4 - 12x - 5 = 0.$
11. $x^4 + 8x^3 + 9x^2 - 8x - 10 = 0.$
12. $x^4 - 2x^2 + 8x - 3 = 0.$
13. $x^4 - 10x^2 - 20x - 16 = 0.$
14. $x^4 - 6x^3 + 3x^2 + 22x - 6 = 0.$
15. $x^4 + 12x - 5 = 0.$
16. $x^4 - 12x + 3 = 0$
17. $x^4 - 8x^2 - 24x + 7 = 0.$
18. $x^4 - 5x^2 - 6x - 5 = 0.$